MARINE GEOLOGY

THE LIVING EARTH

MARINE GEOLOGY

EXPLORING THE NEW FRONTIERS
OF THE OCEAN
REVISED EDITION

JON ERICKSON
FOREWORD BY TIMOTHY KUSKY, Ph.D.

® Checkmark Books®
An imprint of Facts On File, Inc.

MARINE GEOLOGY
Exploring the New Frontiers of the Ocean, Revised Edition

Checkmark Books
An imprint of Facts On File, Inc.
132 West 31st Street
New York NY 10001

Library of Congress Cataloging-in-Publication Data

Erickson, Jon, 1948–
    Marine geology: exploring the new frontiers of the ocean/Jon Erickson.—Rev. ed.
        p. cm.—(The living earth)
Includes bibliographical references and index.
    ISBN 0-8160-4874-6 (hardcover: alk. paper)   ISBN 0-8160-5077-5 (pbk)
    1. Submarine geology. 2. Marine biology. I. Title.
    QE39 E68 2003
    551.46'08—dc21   2002001295

Checkmark Books are available at special discounts when purchased in bulk quantities for businesses, associations, institutions, or sales promotions. Please call our Special Sales Department in New York at 212/967-8800 or 800/322-8755.

You can find Facts On File on the World Wide Web at http://www.factsonfile.com

Text design by Cathy Rincon
Cover design by Nora Wertz
Illustrations by Jeremy Eagle, © Facts On File

Printed in the United States of America

VB Hermitage 10 9 8 7 6 5 4 3 2 1

This book is printed on acid-free paper.

CONTENTS

TABLES

ACKNOWLEDGMENTS

The author thanks the National Aeronautics and Space Administration (NASA), the National Oceanic and Atmospheric Administration (NOAA), the U.S. Army Corps of Engineers, the U.S. Department of Agriculture-Forest Service, the U.S. Department of Agriculture-Soil Conservation Service, the U.S. Defense Nuclear Agency, the U.S. Department of Energy, the U.S. Geological Survey (USGS), the U.S. Maritime Administration, the U.S. Navy, and the Woods Hole Oceanographic Institution (WHOI) for providing photographs for this book.

The author thanks Frank K. Darmstadt, Senior Editor, and the rest of the Facts On File staff for their invaluable contributions to the making of this book.

FOREWORD

Oceans cover approximately two-thirds of the Earth's surface, yet we have explored less of the ocean's depths and mysteries than the surfaces of several nearby planets. The oceans have inspired myths and legends and have been the sources of intrigue, fear, and hope for thousands of years. They have hindered migration of peoples and biota between distant continents yet paradoxically now serve as a principal means of transportation. Oceans provide us with incredible mineral wealth and renewable food and energy sources yet also breed devastating hurricanes. Life on Earth may have begun in environments around hot volcanic events on the seafloor, and we are only beginning today to explore the diverse and unique fauna that thrive in deep, dark waters around similar vents.

In the revised edition of *Marine Geology,* Jon Erickson explores several ideas hypothesizing the origin of the Earth, continents, and oceans and how these processes fit into the origin of the universe. The role of oceans and water in the development of plate tectonics is discussed in detail, while the reader is given essential information on how plate tectonics works. Ocean basins have continually expanded and contracted on Earth, and the continents have alternately converged into large single supercontinents and then broken apart by the formation of new ocean basins. The appearance, evolution, and extinction of different life-forms are inextricably linked to the expansion and contraction of ocean basins, partly through the changing environmental conditions associated with tectonic processes. The history of several different ocean basins over the past billion years is discussed in *Marine Geology,* as well as the changing life-forms in each successive ocean basin.

Erickson presents a fascinating history of ocean exploration. He shows how early explorations were slowly able to reveal data about ocean currents and routes to distant lands and how some dredging operations uncovered huge deposits of metals on the seafloor. Tremendous leaps in our understanding of the structure and topography of the seafloor were acquired during surveying for the navigation of submarines and detecting enemy submarines during World War II. Magnetometers towed behind ships, and accurate depth soundings provided data that led to the formulation of the hypothesis of seafloor spreading, adding the oceanic counterpart to the idea of continental drift. Together these two theories became united as the plate tectonic revolution. This sets the stage for succeeding chapters on the mid-ocean ridges, deep-sea trenches, and submarine volcanoes.

Ocean circulation is responsible for much of the world's climate. Mild foggy winters in London are caused by warm waters from the Gulf of Mexico flowing across the Atlantic in the Gulf Stream to the coast of the British Isles. Large variations in ocean and atmospheric circulation patterns in the Pacific lead to alternating wet and dry climate conditions known as El Niño and La Niña. These variations affect Pacific regions most strongly but are felt throughout the world.

Other movements of water are more dramatic, including the sometimes devastating tsunami that may be initiated by earthquakes, volcanic eruptions, and giant submarine landslides. One of the most tragic tsunami in recent history was generated by the eruption of the Indonesian volcano Krakatau [Krakatoa] in 1883. When Krakatau erupted, it blasted out a large part of the center of the volcano, and seawater rushed in to fill the hole. This seawater was immediately heated, and it exploded outward in a steam eruption and a huge wave of hot water. The tsunami generated by this eruption reached more than 120 feet in height and killed an estimated 36,500 people in nearby coastal regions. In 1998 a catastrophic 50-foot-high wave unexpectedly struck Papua New Guinea, killing more than 2,000 people and leaving more than 10,000 homeless.

The oceans are full of rich mineral deposits, including oil and gas on the continental shelves and slopes and metalliferous deposits formed near mid-ocean ridge vents. Much of the world's wealth of manganese, copper, and gold lies on the seafloor. The oceans also yield rich harvests of fish, and care must be taken that we do not deplete this source by overfishing. Sea vegetables are growing in popularity and their use may help alleviate the increasing demand for space in fertile farmland. The oceans offer the world a solution to increasing energy and food demands in the face of a growing world population. New life-forms are constantly being discovered in the ocean's depths, and understanding these creatures is necessary before any changes we make to their environment causes them to perish forever.

— Timothy M. Kusky, Ph.D.

INTRODUCTION

This planet contains so much water that perhaps it should have been better named Oceania. It is the only known body in the solar system that is surrounded by water filled with unique geologic structures and teeming with a staggering assortment of marine life. Some of the strangest creatures on Earth, whose ancestors go back several hundred million years, live on the deep ocean floor. Many undersea ridges host an eerie world that time forgot—a cold, dark abyss consisting of tall chimneys spewing hot, mineral-rich water that supports unusual species previously unknown to science.

The floor of the ocean presents a rugged landscape unmatched anywhere on the continents. Vast undersea mountain ranges much more extensive than those on land crisscross the seabed. Although deeply submerged, the midocean ridges are easily the most prominent features on the planet. The ocean floor is continuously being created at spreading ridges, where molten rock oozes out of the mantle, and destroyed in the deepest trenches of the world. Much of the world's untapped wealth lies undersea. The seabed therefore offers new frontiers for future exploration of mineral resources.

An extraordinary number of volcanoes are hidden under the waves, many more than on the land. Most of the volcanic activity that continually remakes the surface of Earth occurs on the ocean floor. Active volcanoes rising up from the bottom of the ocean create the tallest mountains. Most of the world's islands in fact began as undersea volcanoes that broke the surface of the sea. However, the preponderance of marine volcanoes is not exposed at the surface but spread out on the ocean floor as isolated seamounts.

Chasms that challenge the largest terrestrial canyons plunge to great depths. Massive submarine slides gouge deep depressions into the seabed and deposit enormous heaps of sediment onto the ocean floor. Undersea slides also occasionally generate tall waves that pound nearby shores, causing much destruction to seaside communities. Abyssal storms with strong currents sculpt the ocean bottom, churning up huge clouds of sediment and dramatically modifying the seafloor. The scouring of the seabed and deposition of large amounts of sediment result in a highly complex marine geology.

This revised and updated edition is a much expanded and more inclusive examination of the intriguing subject of marine geology. Science enthusiasts will particularly enjoy this fascinating subject and gain a better understanding of how the forces of nature operate on Earth. Students of geology and Earth science will also find this a valuable reference book to further their studies. Readers will enjoy this clear and easily readable text that is well illustrated with extraordinary photographs, detailed illustrations, and helpful tables. A comprehensive glossary is provided to define difficult terms, and a bibliography lists references for further reading. The geologic processes that shape the surface of this planet are an example of the spectacular forces that create the living Earth.

1

THE BLUE PLANET
THE WORLD'S OCEANS

This opening chapter chronicles the formation of Earth and the evolution of the oceans. Earth is unique among planets, because it is the only body in the solar system with a water ocean and an oxygen atmosphere. As many as 20 oceans have come and gone throughout this planet's long history (Table 1) as continents drifted apart from each other and reconverged into supercontinents. The present oceans formed when a supercontinent named Pangaea, Greek for "all lands," broke apart into today's continents about 170 million years ago.

Prior to the breakup of Pangaea, a single large ocean called Panthalassa, Greek for "universal sea," surrounded the supercontinent. Before the assemblage of Pangaea, all continents surrounded an ancient Atlantic Ocean called the Iapetus Sea. Deeper into the past, the continents again formed a supercontinent named Rodinia, Russian for "Motherland." Its breakup created entirely new seas, which participated in life's greatest explosion of new species. Life itself possibly evolved at the bottom of a global ocean not long after Earth's creation.

TABLE 1   THE GEOLOGIC TIME SCALE

Era	Period	Epoch	Age (millions of years)	First Life-forms	Geology
		Holocene	0.01		
	Quaternary				
		Pleistocene	3	Humans	Ice age
Cenozoic		Pliocene	11	Mastodons	Cascades
		Neogene			
		Miocene	26	Saber-toothed tigers	Alps
	Tertiary	Oligocene	37		
		Paleogene			
		Eocene	54	Whales	
		Paleocene	65	Horses, Alligators	Rockies
	Cretaceous		135		
				Birds	Sierra Nevada
Mesozoic	Jurassic		210	Mammals	Atlantic
				Dinosaurs	
	Triassic		250		
	Permian		280	Reptiles	Appalachians
	Pennsylvanian		310		Ice age
				Trees	
	Carboniferous				
Paleozoic	Mississippian		345	Amphibians	Pangaea
				Insects	
	Devonian		400	Sharks	
	Silurian		435	Land plants	Laursia
	Ordovician		500	Fish	
	Cambrian		570	Sea plants	Gondwana
				Shelled animals	
			700	Invertebrates	
Proterozoic			2500	Metazoans	
			3500	Earliest life	
Archean			4000		Oldest rocks
			4600		Meteorites

ORIGIN OF SEA AND SKY

An incredible amount of water resides in the solar system, much more than on Earth alone. As the Sun emerged from gas and dust, tiny bits of ice and rock debris gathered in a frigid, flattened disk of planetesimals surrounding the infant star. The temperatures in some parts of the disk might have been warm enough for liquid water to exist on the first solid bodies to form. In addition, water vapor in the primordial atmospheres of the inner terrestrial planets might have been eroded away by planetesimal bombardment and blown beyond Mars by the strong solar wind of the infant Sun. Once planted in the far reaches of the solar system, ice crystals coalesced into comets that returned to Earth to supply it with additional water.

The creation of the Moon (Fig. 1) remains a mystery. Perhaps a Mars-sized body slammed into Earth and splashed enough material into orbit to coalesce into a daughter planet. The presence of a rather large moon, the biggest of any moon in the solar system in relation to its mother planet, might have had a major influence on the initiation of life. The unique properties of the Earth-Moon system raised tides in the ocean. Cycles of wetting and drying in tidal pools might have helped the planet acquire life much earlier than previously thought possible. The Moon might also have been responsible for the relatively

Figure 1 *The surface of the Moon viewed from an Apollo spacecraft showing many of its terrain features.*

(Photo courtesy NASA)

Figure 2 *Zircons from the rare-earth zone, Jasper cuts area, Gilpin County, Colorado.*

(Photo by E. J. Young, courtesy USGS)

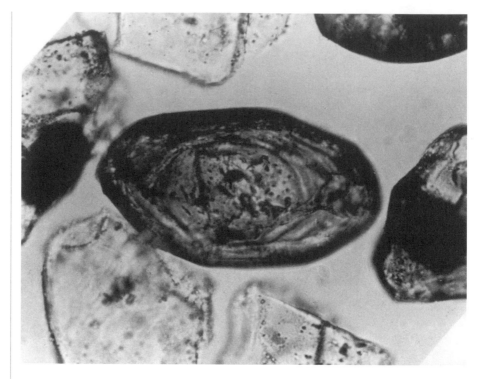

stable climate. It might have made Earth hospitable to life by stabilizing the tilt of the planet's rotational axis, which marks the seasons. Without the Moon, life on Earth would likely face the same type of wild fluctuations in climate that Mars has apparently experienced through the eons.

Earth's original crust was quite distinct from modern continental crust, which first appeared about 4 billion years ago and represents less than 0.5 percent of the planet's total volume. During this time, Earth spun wildly on its axis, completing a single rotation every 14 hours, thus maintaining high temperatures throughout the planet. Present-day plate tectonic processes could not have operated under such hot conditions, which produced more vertical bubbling than horizontal sliding. Therefore, modern-style plate tectonic processes were probably not fully functional until 2.7 billion years ago, when the formation of the crust was nearly complete.

Earth apparently took less than half its history to form an equivalent volume of continental rock it has today. Information about the early crust is provided by some of the most ancient rocks that survived intact. They formed deep within the crust a few hundred million years after the formation of the planet and now outcrop at the surface. Zircon crystals (Fig. 2) found in granite are enormously resistant and tell of the earliest history of Earth, when the crust first formed some 4.2 billion years ago. Among the oldest rocks is the 4-billion-year-old Acasta gneiss, a metamorphosed granite in the Northwest Territories of Canada. Its existence suggests that the formation of the crust was well underway by this time. The discovery is used as evidence that at least small patches of continental crust existed on Earth's surface at an early date.

During Earth's formative years, a barrage of asteroids and comets pounded the infant planet and the Moon between 4.2 and 3.9 billion years ago. A swarm of debris left over from the creation of the solar system bombarded Earth. The bombardment might have delivered heat and organic compounds to the planet, sparking the rapid formation of primitive life. Such a pummeling could also have wiped out existing life-forms in a colossal mass extinction.

Comets comprising rock debris and ice also plunged into Earth, releasing tremendous quantities of water vapor and gases. The degassing of these cosmic invaders produced mostly carbon dioxide, ammonia, and methane, major constituents of the early atmosphere, which began to form about 4.4 billion years ago. Most of the water vapor and gases originated from within Earth itself by volcanic outgassing. The early volcanoes were extremely explosive because Earth's interior was hotter and the magma contained more abundant volatiles than today.

Earth soon acquired a primordial atmosphere composed of carbon dioxide, nitrogen, water vapor, and other gases belched from a profusion of volcanoes. Water vapor so saturated the air that atmospheric pressure was several times greater than it is today. The early atmosphere contained up to 1,000 times the current level of carbon dioxide. This was fortunate because the Sun's output was only about 75 percent its present value, and a strong greenhouse effect kept the oceans from freezing solid. The planet also retained heat by a high rotation rate, with days only 14 hours long, and by the absence of continents to block the flow of ocean currents.

Oxygen originated directly from volcanic outgassing and meteorite degassing. It also developed indirectly from the breakdown of water vapor and carbon dioxide by strong ultraviolet radiation from the Sun. All oxygen generated by these methods quickly bonded to metals in the crust, much like the rusting of iron. Oxygen also recombined with hydrogen and carbon monoxide to reconstitute water vapor and carbon dioxide. A small amount of oxygen existing in the upper atmosphere might have provided a thin ozone screen. This would have reduced the breakdown of water molecules by strong ultraviolet rays from the Sun, thus preventing the loss of the entire ocean, a fate that might have visited Venus eons ago (Fig. 3).

Nitrogen originated from volcanic eruptions and from the breakdown of ammonia. The ammonium molecule, composed of one nitrogen atom and three hydrogen atoms, is also a major constituent of the primordial atmosphere. Unlike most other gases, which have been replaced or recycled, Earth

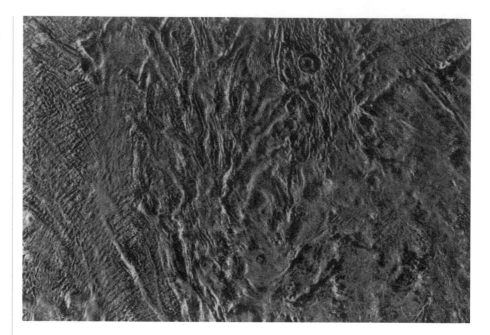

retains much of its original nitrogen. This is because nitrogen readily transforms into nitrate, which easily dissolves in the ocean. As a result, the denitrifying bacteria return the nitrate–nitrogen to its gaseous state. Therefore, without life, Earth would have long ago lost its nitrogen and possess only a fraction of its present atmospheric pressure.

The surface of Earth was scorching hot and in a constant rage. Winds blew with tornadic force, producing fierce dust storms that raced across the dry surface. The planet was blanketed with suspended sediment similar to Martian dust storms today (Fig. 4). Huge lightning bolts darted back and forth. Earth-shattering thunder sent gigantic shock waves reverberating through the air. Volcanoes erupted in one gigantic outburst after another, lighting the sky with white-hot sparks of ash and red-hot lava flowing across the land.

The restless Earth rent apart as massive quakes cracked open the thin crust and huge batches of magma bled through the fissures. Voluminous lava flows flooded the surface, forming flat, featureless plains dotted with towering volcanoes. The intense volcanism also lofted massive quantities of volcanic debris high into the atmosphere, giving the sky an eerie red glow. The dust also cooled the planet and acted as particles upon which water vapor coalesced.

With a further drop in atmospheric temperatures, water vapor condensed into heavy clouds that shrouded the planet. They completely blocked out the Sun and plunged the surface into total darkness. As the atmosphere continued to cool, sheets of rain fell from the sky. Deluge after deluge overflowed the landscape. Raging floods cascaded down steep volcanic slopes and the sides of large meteorite craters, gouging out deep ravines in the rocky plain.

Around 4 billion years ago, when the rains ceased and the skies finally cleared, Earth emerged as a giant blue orb covered by a global ocean nearly 2 miles deep dotted with numerous volcanic islands. Countless volcanoes erupted undersea. Hydrothermal vents spewed hot water containing sulfur and other chemicals. Some volcanoes poked through the sea surface to emerge as scattered islands, yet no continents graced Earth's watery face.

Massive meteorite impacts continued to strike Earth, making living conditions very difficult for life, which was first striving to organize proteins into living cells. The first cells might have been repeatedly exterminated, forcing life to originate again and again. Whenever primitive organic molecules attempted to arrange themselves into living matter, frequent impacts blasted them apart before they had a chance to reproduce.

Some large impactors might have generated enough heat to boil off a large portion of the ocean repeatedly. The vaporized sea would have raised surface pressures more than 100 atmospheres. The resulting high temperatures would have sterilized the entire planet. Several thousand years elapsed before Earth cooled sufficiently for steam to condense into rain and refill the ocean basins once again, only to await the next ocean-evaporating impact. Such harsh conditions could have set back the emergence of life hundreds of millions of years.

Perhaps the only safe place for life to evolve was on the deep ocean floor, where a high density of hydrothermal vents existed. These vents acted like geysers on the bottom of the sea (Fig. 5) and expelled mineral-laden hot water heated by shallow magma chambers resting just beneath the ocean floor. The vents might have created an environment capable of generating an immense number of organic reactions and providing the ingredients and energy necessary to create the planet's first life. The vents would have also given evolving

Figure 4 The surface of Mars view from Viking Lander 2, *showing a boulder-strewn field with rocks embedded in fine sediment from Martian dust storms.*

(Photo courtesy NASA)

7

Figure 5 *Hydrothermal*
vents on the deep ocean
floor provide nourishment
and heat for bottom
dwellers.

(Photo courtesy USGS)

life-forms all the essential nutrients needed to sustain themselves. Indeed, such an environment exists today and is home to some of the strangest creatures found on Earth. In this environment, life could have originated as early as 4.2 billion years ago.

Evidence for life early in Earth's history, when the planet was still quite hot, exists today as thermophilic (heat-loving) bacteria. They are found in thermal springs and other hot-water environments throughout the world (Fig. 6) as well as deep underground or far beneath the ocean floor. Archaebacteria, or simply archaea, range widely throughout the seas. Many parts of the ocean are teeming with them. One-third of the biomass of picoplankton (the tiniest plankton) in Antarctic waters comprises archaea. Such abundance could mean that archaea play an important role in the global ecology and might have a major influence on the chemistry of the ocean.

THE UNIVERSAL SEA

The early Earth was mostly ocean. Continental crust comprised slivers of granite that drifted freely over the surface and was perhaps only one-tenth its current size. Between the end of the great meteorite bombardment and the formation of the first sedimentary rocks about 3.8 billion years ago, large volumes of water flooded Earth's surface. Seawater probably began salty due to the volcanic outgassing of chlorine and sodium. However, the ocean did not contain its present concentration of salts until about 500 million years ago. The salt level has remained generally constant ever since. Yet the ocean has not always been consistent. Major changes in seawater chemistry often correlate with biologic extinction (Table 2).

In a remote mountainous area in southwest Greenland, metamorphosed marine sediments of the Isua Formation provide strong evidence for an early saltwater ocean. The Isua rocks originated in volcanic island arcs and therefore lend credence to the idea that plate tectonics operated early in the history of Earth. The rocks are among the most ancient, dating to about 3.8 billion years, and indicate that the planet had abundant surface water by this time. Earth's oldest sedimentary rocks also contain chemical fingerprints of complex cells that originated as early as 3.9 billion years ago.

Most of the crust was deeply submerged during the Archean, as evidenced by an abundance of chert deposits more than 2.5 billion years old. Chert is among the hardest minerals. Some appears to have precipitated from silica–rich water in deep oceans. The seas contained abundant dissolved silica, which leached out of volcanic rock pouring onto the ocean floor. Modern seawater is deficient in silica because organisms such as diatoms extract it to build ornate skeletons (Fig. 7). Massive deposits of diatomaceous earth, or diatomite, are a tribute to the great success of these organisms during the last 600 million years.

Figure 6   *Boiling mud springs northwest of Imperial Junction, California.*

(Photo by W. C. Mendenhall, courtesy USGS)

TABLE 2   RADIATION AND EXTINCTION OF SPECIES

Organism	Radiation	Extinction
Mammals	Paleocene	Pleistocene
Reptiles	Permian	Upper Cretaceous
Amphibians	Pennsylvanian	Permian-Triassic
Insects	Upper Paleozoic	
Land plants	Devonian	Permian
Fish	Devonian	Pennsylvanian
Crinoids	Ordovician	Upper Permian
Trilobites	Cambrian	Carboniferous & Permian
Ammonoids	Devonian	Upper Cretaceous
Nautiloids	Ordovician	Mississippian
Brachiopods	Ordovician	Devonian & Carboniferous
Graptolites	Ordovician	Silurian & Devonian
Foraminiferans	Silurian	Permian & Triassic
Marine invertebrates	Lower Paleozoic	Permian

Sulfur from volcanoes was particularly abundant in the early ocean and combined easily with metals such as iron to form sulfates. Biochemical activity in the ocean was also responsible for stratified sulfide deposits. Sulfur-metabolizing bacteria living near undersea hydrothermal vents oxidized hydrogen sulfide into elemental sulfur and various sulfates. Copper, lead, and zinc, which were much more abundant in the Proterozoic than in the Archean, also reflect a submarine volcanic origin.

The earliest organisms were sulfur-metabolizing bacteria similar to those living symbiotically in the tissues of tube worms near sulfurous hydrothermal vents on the East Pacific Rise and on a dozen other midocean ridges scattered around the world. Abundant sulfur in the early sea provided the nutrients to sustain life without the need for oxygen. Bacteria obtained energy by the reduction (opposite of oxidation) of this important element. The growth of primitive bacteria was also limited by the amount of organic molecules produced in the ocean.

Oxygen was practically nonexistent when life first appeared (Table 3). Oxygen levels remained low because of the oxidation of dissolved metals in seawater and reduced gases emitted from submarine hydrothermal vents. The seas contained abundant iron, which reacted with oxygen generated by photosynthesis. This was a fortunate circumstance since oxygen was also poisonous to primitive life-forms. The oxygen level jumped significantly between 2.2 and 2 billion years ago. During that time, the ocean's high concentration of dissolved iron, which depleted free oxygen to form iron oxide, was deposited onto the seafloor, creating the world's great iron ore reserves.

Iron leached from the continents and dissolved in seawater reacted with oxygen in the ocean and precipitated in massive ore deposits on shallow continental margins. The average ocean temperature was much warmer than it is

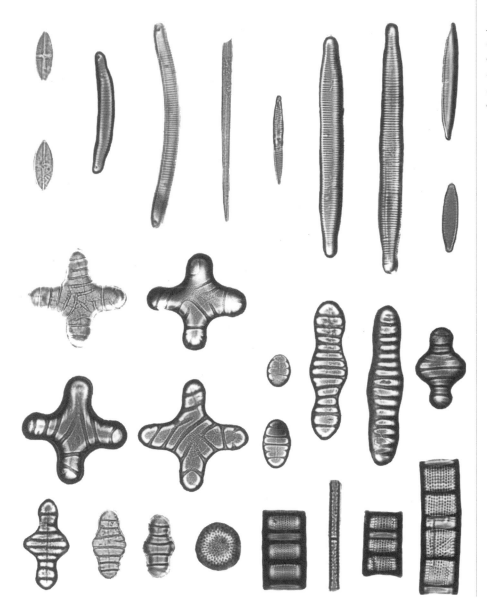

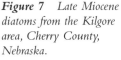

Figure 7 *Late Miocene diatoms from the Kilgore area, Cherry County, Nebraska.*

(Photo by G. W. Andrews, courtesy USGS)

TABLE 3   EVOLUTION OF THE BIOSPHERE

	Billions of Years Ago	Percent Oxygen	Biologic Effects	Event Results
Full oxygen conditions	0.4	100	Fish, land plants, and animals	Approach present biologic environs
Appearance of shell-covered animals	0.6	10	Cambrian fauna	Burrowing habitats
Metazoans appear	0.7	7	Ediacaran fauna	First metazoan fossils and tracks
Eukaryotic cells appear	1.4	>1	Larger cells with a nucleus	Redbeds, multicellular organisms
Blue-green algae	2.0	1	Algal filaments	Oxygen metabolism
Algal precursors	2.8	<1	Stromatolite mounds	Initial photosynthesis
Origin of life	4.0	0	Light carbon	Evolution of the biosphere

today. When warm ocean currents rich in iron and silica flowed toward the polar regions, the suddenly cooled waters could no longer hold minerals in solution. The minerals consequently precipitated out of seawater. They formed alternating layers due to the difference in settling rates between silica and iron, the heavier of the two minerals.

A drop in global temperatures initiated the first known glaciation about 2.4 billion years ago (Table 4) when massive ice sheets nearly engulfed the entire landmass. The positions of the continents also had a tremendous influence on the initiation of ice ages. When lands wandered into the colder latitudes, glacial ice built up. Global tectonics, featuring extensive volcanic activity and seafloor spreading, might have triggered the ice ages by drawing down the level of oxygen in the ocean and atmosphere.

The loss of oxygen preserved more organic carbon in the sediments, preventing living organisms from returning it to the atmosphere. The elimination of carbon dioxide in this manner caused Earth to cool dramatically. Besides high rates of organic carbon burial, iron deposition and intense hydrothermal activity associated with plate tectonics furthered global cooling. Although, this was the first ice age the world had ever known, it was not the worst.

The burial of carbon in Earth's crust might have been the key to the onset of another glacial period just before the appearance of recognizable animal life about 680 million years ago. This is called the Varanger ice age, named for a Fjord in Norway. Massive glaciers overran nearly half the continents for millions of years. Four periods of glaciation occurred between 850 and 580 million years ago, when even the Tropics froze over. The intensity of the glaciation is indicated by the occurrence of glacial debris and unusual deposits of iron-rich rocks on virtually every continent, indicating they formed in cold waters.

It was perhaps the greatest and most prolonged period of glaciation when ice encased nearly half the planet. The climate was so cold that ice sheets and permafrost (permanently frozen ground) extended into equatorial latitudes. Ice cover was so extensive that the period has been dubbed the "snowball Earth." If not for massive volcanic activity, which restored the carbon dioxide content of the atmosphere by the greenhouse effect, the planet could still be buried under ice.

During this time, only single-celled plants and animals lived in the sea. This ice age dealt a deathly blow to life in the ocean, and many simple organisms vanished during the world's first mass extinction. At this point in Earth's development, animal life was still scarce. The extinction decimated the ocean's population of acritarchs, a community of planktonic algae that were among the first organisms to develop elaborate cells with nuclei. These extreme glaciations took place just before a rapid diversification of multicellular life, culminating in an explosion of new species (Fig. 8).

TABLE 4   THE MAJOR ICE AGES

Number of years ago	Event
10,000–present	Present interglacial
15,000–10,000	Melting of ice sheets
20,000–18,000	Last glacial maximum
100,000	Most recent glacial episode
1 million	First major interglacial
3 million	First glacial episode in Northern Hemisphere
4 million	Ice covers Greenland and the Arctic Ocean
15 million	Second major glacial episode in Antarctica
30 million	First major glacial episode in Antarctica
65 million	Climate deteriorates, poles become much colder
250–65 million	Interval of warm and relatively uniform climate
250 million	The great Permian ice age
700 million	The great Precambrian ice age
2.4 billion	First major ice age

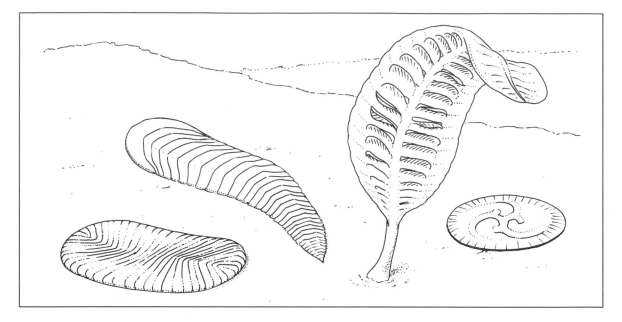

Figure 8 *The late Precambrian Ediacaran fauna from Australia.*

About 2 billion years ago, after most of the dissolved iron was locked up in the sediments, the level of oxygen began to rise and replace carbon dioxide in the ocean and atmosphere. Major plate tectonic cataclysms caused oxygen levels to surge between 2.1 and 1.7 million years ago and again between 1.1 and 0.7 billion years ago. This occurred during the breakup of supercontinents and the formation of new ocean basins.

The best evidence for plate tectonics operating early in Earth's history is found in greenstone belts scattered in various parts of the world. They are a mixture of metamorphosed (recrystallized) lava flows and sediments possibly derived from island arcs (chains of volcanic islands on the edges of deep-sea trenches) caught between colliding continents. Although no large continents existed during this time, the foundations upon which they formed were present as protocontinents. These small landmasses were separated by marine basins that accumulated lava and sediments derived mainly from volcanic rocks that later metamorphosed into greenstone belts.

Ophiolites, from the Greek word *ophis* meaning "serpent," captured in greenstones are slices of ocean floor shoved up onto the continents by drifting plates and date as old as 3.6 billion years. Blueschists (Fig. 9) are metamorphosed rocks of subducted ocean crust thrust onto the continents by plate motions. Pillow lavas, which are tubular bodies of basalt extruded undersea, also appear in the greenstone belts, signifying that the volcanic eruptions took place on the ocean floor.

Toward the end of the Precambrian, some 700 million years ago, all land-masses assembled into the supercontinent Rodinia (Fig. 10), centered over the equator. The continental collision resulted in environmental changes that had a profound influence on the evolution of life. The ancient North American continent called Laurentia lay at the core of Rodinia, while Australia and Antarctica bordered on its west coast. Laurentia, comprising the interior of North America, Greenland, and northern Europe, assembled during the collision of several microcontinents beginning about 1.8 billion years ago and evolved in a relatively brief period of only 150 million years.

Laurentia continued to grow by garnering bits and pieces of continents and chains of young volcanic islands. At Cape Smith on the Hudson Bay, in Quebec, Canada, lies a piece of oceanic crust that was squeezed onto the land some 1.8 billion years ago. This is a telltale sign that continents collided and closed off an ancient sea. Arcs of volcanic rock also weave through central and eastern Canada down into the Dakotas. About 700 million years ago, Laurentia collided with another large continent on its southern and eastern edges, creating the supercontinent. A superocean positioned in the approximate location of today's Pacific Ocean surrounded the supercontinent. No broad oceans or extreme differences in temperature existed to prevent species from migrating to various parts of the world.

Between 630 and 560 million years ago, the supercontinent rifted apart, and four or five continents rapidly drifted away from each other. As the con-

Figure 9   *An outcrop of retrograde blueschist rocks in the Seward Peninsula region, Alaska.*

(Photo by C. L. Sainsbury, courtesy USGS)

Figure 10 *The supercontinent Rodinia, 700 million years ago.*

tinents dispersed and subsided, seas flooded their interiors, creating large continental shelves. Most continents huddled near the equator, which explains the presence of warm Cambrian seas. The continental breakup caused sea levels to rise and flood large portions of the land at the beginning of the Cambrian. This setting heralded an explosion of new life-forms, the likes of which have never been seen before or since (Fig. 11).

THE IAPETUS SEA

When Rodinia rifted apart, the separated continents opened a proto-Atlantic Ocean called the Iapetus (Fig. 12). The rifting process formed extensive inland seas that submerged most of Laurentia some 540 million years ago, as evidenced by the presence of Cambrian seashores in such places as Wisconsin. It also flooded the ancient European continent called Baltica. In the Southern Hemisphere, continental motions assembled the present continents of South

America, Africa, Australia, Antarctica, and India into the supercontinent Gondwana (Fig. 13), named for a geologic province in east-central India.

The formation of the Iapetus created extensive inland seas that inundated most of Laurentia and Baltica. The Iapetus Sea was similar in size to the present-day North Atlantic and occupied the same general location. A continuous coastline running from Georgia to Newfoundland between about 570 and 480 million years ago suggests that this ancient east coast faced a wide, deep sea that stretched at least 1,000 miles across from east to west and bordered a much larger body of water to the south.

Volcanic islands dotted the Iapetus Sea, which resembled the present-day Pacific Ocean between Southeast Asia and Australia. About 460 million years ago, the shallow waters of the nearshore environment of this ancient sea contained abundant invertebrates, including prolific trilobites (Fig. 14). These were small, oval arthropods that accounted for about 70 percent of all species and are a favorite among fossil collectors. Eventually, the trilobites faded, while mollusks and other invertebrates took their place and expanded throughout the seas.

Between about 420 million and 380 million years ago, Laurentia collided with Baltica, closing off the Iapetus. The collision fused the two continents into Laurasia, named for the Laurentian province of Canada and the Eurasian continent. The Eurasian continent, the largest landmass in the world, formed when some dozen individual continental blocks welded together about a half-billion years ago.

Figure 11 *Early Cambrian marine fauna.*

Figure 12 *About 500 million years ago, the continents surrounded an ancient sea called the Iapetus.*

Figure 13 *The configuration of the southern continents that comprised Gondwana.*

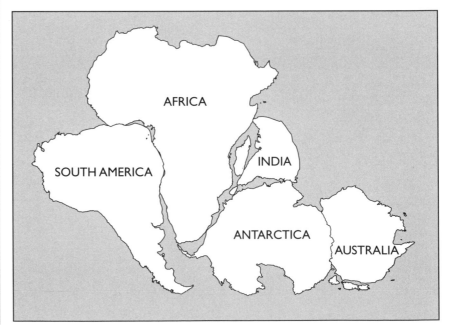

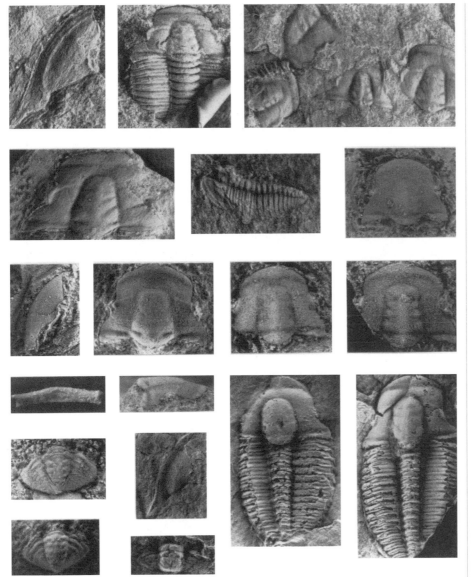

Figure 14 *Fossil
trilobites of the Carrara
Formation in the southern
Great Basin, California-
Nevada.*

(Photo by A. R. Palmer,
courtesy USGS)

The closing of the Iapetus resulted in large-scale mountain building. The continents flanking the sea collided and pushed up great mounds of crust in northern Europe and North America, including those that evolved into the Appalachians. The spate of mountain building might have triggered a burst of species diversity. The largest of these increases was the radiation of marine species around 450 million years ago. The foreland basins filled with thick sediments eroded from nearby mountains. Erosion of the mountains might have pumped nutrients into the sea, fueling booms of marine plankton, which increased the food supply for higher creatures. The number of genera of mollusks, brachiopods, and trilobites dramatically increased, because organisms with abundant food are more likely to thrive and diversify into different species.

During the formation of Laurasia, island arcs between the two landmasses were scooped up and plastered against continental edges as the oceanic crustal plate carrying the islands subducted under Baltica. This subduction rafted the islands into collision with the continent and deposited the formerly submerged rocks onto the present west coast of Norway. Slices of land called *terranes* residing in western Europe drifted into the Iapetus from ancient Africa. Likewise, slivers of crust from Asia traveled across the ancestral Pacific Ocean called the Panthalassa to form much of western North America.

North America was a lost continent around 500 million years ago. During that time, the continental landmass and a few smaller continental fragments drifted freely on their own. South America, Africa, Australia, Antarctica, and India had assembled into Gondwana by continental plate collisions. At this time, North America was situated a few thousand miles off the western coast of South America, placing it on the western side of Gondwana. Eventually, North and South America collided (Fig. 15), placing what would be present-day Washington, D.C., near Lima, Peru. A limestone formation in Argentina

Figure 15 *North and South America might have collided at the beginning of the Ordovician 500 million years ago.*

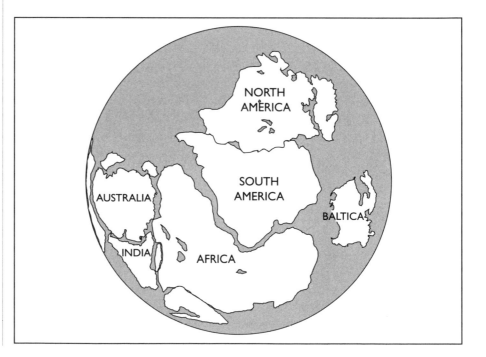

contains a distinctive trilobite species typical of North America but not of South America, suggesting the two continents once had much in common.

THE PANTHALASSA SEA

Throughout geologic history, smaller continental blocks collided and merged into larger continents. Millions of years after assembling, the continents rifted apart, and the chasms filled with seawater to form new oceans. However, the regions presently bordering the Pacific basin apparently did not collide. Rather, the Pacific Ocean is a remnant of an ancient sea called the Panthalassa. It narrowed and widened in response to continental breakup, dispersal, and reconvergence in the area occupied by today's Atlantic Ocean. So, while oceans have repeatedly opened and closed in the vicinity of the Atlantic basin, a single ocean has existed continuously at the site of the Pacific basin.

When Laurentia fused with Baltica to form Laurasia, island arcs in the Panthalassa Sea began colliding with the western margin of present-day North America. Erosion leveled the continents. Shallow seas flowed inland, flooding more than half the land surface. The inland seas and wide continental margins, along with a stable environment, encouraged marine life to flourish and spread throughout the world.

From 360 million to 270 million years ago, Gondwana and Laurasia converged into Pangaea (Fig. 16), which straddled the equator and extended almost from pole to pole. This massive continent reached its peak size about 210 million years ago with an area of about 80 million square miles or 40 percent of Earth's total surface area. More than one-third of the landmass was covered with water. An almost equal amount of land existed in both hemispheres. In contrast, today two-thirds of the continental landmass is located north of the equator. South of the equator, the breakdown is 10 percent landmass and 90 percent ocean. A single great ocean stretched uninterrupted across the planet, while the continents huddled to one side of the globe.

The sea level fell substantially after the formation of Pangaea, draining the interiors of the continents and causing the inland seas to retreat. A continuous shallow-water margin ran around the entire perimeter of Pangaea. As a result, no major physical barriers hampered the dispersal of marine life. Moreover, the seas were largely restricted to the ocean basins, leaving the continental shelves mostly exposed.

The continental margins were less extensive and narrower than they are today due to a drop in sea level as much as 500 feet. This drop confined marine habitats to the nearshore regions. Consequently, habitat areas for shallow-water marine organisms were limited, resulting in low species diversity. Permian ocean life was sparse, with many immobile animals and few active

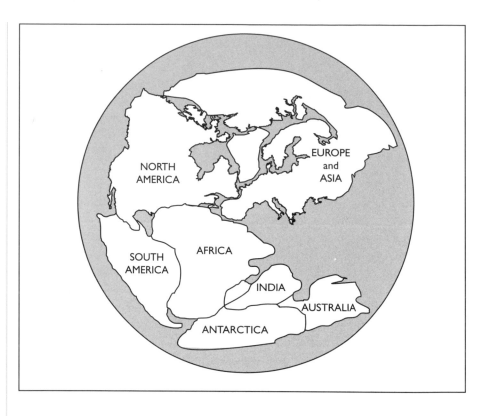

predators. Ocean temperatures remained cool following a late Permian ice age. Marine invertebrates that managed to escape extinction lived in a narrow margin near the equator.

THE TETHYS SEA

When Laurasia occupied the Northern Hemisphere and its counterpart Gondwana was located in the Southern Hemisphere, the two landmasses were separated by a large shallow equatorial body of water called the Tethys Sea (Fig. 17) that was named for the mother of the seas in Greek mythology. After the assembly of Pangaea, the Tethys became a huge embayment separating the northern and southern arms of the supercontinent, which resembled a gigantic letter *C* straddling the equator.

The Tethys was a broad tropical seaway extending from western Europe to southeast Asia that harbored diverse and abundant shallow-water marine life. Reef building in the Tethys Sea was intense, forming thick deposits of limestone and dolomite laid down by prolific lime-secreting organisms. The tropics served as an evolutionary cradle. This is because they had a greater area of shallow seas than other regions, providing an exceptional environment for new organisms to evolve.

During the Mesozoic era, an interior sea flowed into the west-central portions of North America and inundated the area that now comprises eastern Mexico, southern Texas, and Louisiana. A shallow body of water called the Western Interior Cretaceous Seaway (Fig. 18) divided the North American continent into the western highlands, comprising the newly forming Rocky Mountains and isolated volcanoes, and the eastern uplands, consisting of the Appalachian Mountains. Seas also invaded South America, Africa, Asia, and Australia. The continents were flatter, mountain ranges were lower, and sea levels were higher than at present. Thick beds of limestone and dolomite were deposited in the interior seas of Europe and Asia. These rocks later uplifted to form the Alps and Himalayas.

At the beginning of the Cenozoic era, high sea levels continued to flood continental margins and formed great inland seas, some of which split continents in half. Seas divided North America in the Rocky Mountain and high plains regions. South America was cut in two in the region that later became the Amazon basin. Additionally, the joining of the Tethys Sea and the newly

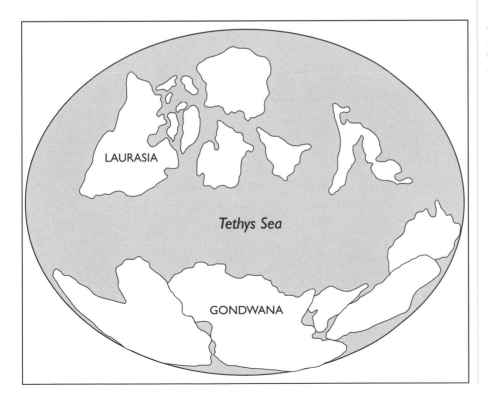

Figure 17 *About 400 million years ago, the continents surrounded an ancient sea called the Tethys.*

Figure 18 *The paleogeography of the Cretaceous period, showing the interior sea.*

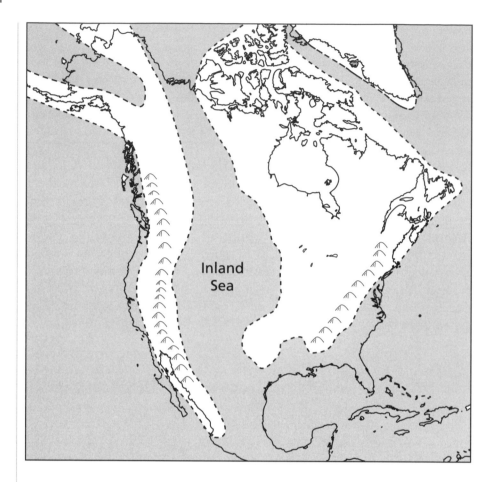

formed Arctic Ocean split Eurasia. The oceans were interconnected in the equatorial regions by the Tethys and Central American seaways. This provided a unique circumglobal oceanic current system that distributed heat to all parts of the world and maintained an unusually warm climate. The higher sea levels reduced the total land surface to perhaps half its present size.

During the Cretaceous period, plants and animals were especially prolific and ranged practically from pole to pole. The deep ocean waters, which are now near freezing, were about 15 degrees Celsius during the Cretaceous. The average global surface temperature was 10 to 15 degrees warmer than at present. Conditions were also much warmer in the polar regions. The temperature difference between the poles and the equator was only 20 degrees, or about half that of today.

The movement of the continents was more rapid than at present, with perhaps the most vigorous plate tectonics the world has ever known. The drifting of continents into warmer equatorial waters might have accounted for much of the mild climate during the Cretaceous. By the time of the initial breakup of the continents about 170 million years ago, the climate began to warm dramatically. The continents were flatter, the mountains were lower, and the sea levels were higher. Although the geography during this time was important, it did not account for all of the warming.

About 120 million years ago, an extraordinary burst of submarine volcanism struck the Pacific basin, releasing vast amounts of greenhouse gas–laden lava onto the ocean floor. The surge of volcanism increased the production of oceanic crust by as much as 50 percent. The amount of atmospheric carbon dioxide rose 10 times the level of today. The volcanic spasm is evidenced by a collection of massive undersea lava plateaus that formed almost simultaneously. The largest of which, the Ontong Java, is about two-thirds the size of Australia. It contains at least 9 million cubic miles of basalt, enough to bury the entire United States beneath 3 miles of lava.

During the final stages of the Cretaceous, the seas receded from the land as sea levels dropped and temperatures in the Tethys Sea began to fall. Most warmth-loving species, especially those living in the tropical Tethys Sea, disappeared when the Cretaceous ended. The most temperature-sensitive Tethyan faunas suffered the heaviest extinction rates. Species that were amazingly successful in the warm waters of the Tethys dramatically declined when ocean temperatures dropped.

Major marine groups that disappeared at the end of the Cretaceous included marine dinosaurs, the ammonoids (Fig. 19), which were ancestors of the nautilus, the rudists, which were huge coral-shaped clams, and other types of clams and oysters. All the shelled cephalopods were absent in the Cenozoic seas except the nautilus and shell-less species, including cuttlefish, octopus, and squid. The squid competed directly with fish, which were little affected by the extinction.

Marine species that survived the great die-off were much the same as those of the Mesozoic era. The ocean has a moderating effect on evolutionary processes because it has a longer "memory" of environmental conditions than does the land, taking much longer to heat up or cool down. Species that inhabited unstable environments, such as those regions in the higher latitudes, were especially successful. Offshore species fared much better than those living in the turbulent inshore waters.

Because of high evaporation rates and low rainfall, warm water in the Tethys Sea became top-heavy with salt and sank to the ocean bottom. Meanwhile, ancient Antarctica, whose climate was warmer than at present, generated cool water that filled the upper layers. This action caused the deep ocean to run backward, circulating from the tropics to the poles, just the opposite of today's patterns. About 28 million years ago, Africa collided with Eurasia and blocked warm water from flowing to the poles, thereby allowing a major ice

Figure 19 *A collection of ammonoid fossils.*

(Photo by M. Gordon Jr., courtesy USGS)

sheet to form on Antarctica. Ice flowing into the surrounding sea cooled the surface waters, which sank to the ocean depths and flowed toward the equator, generating the present-day ocean circulation system.

About 50 million years ago, the Tethys Sea narrowed as the African and Eurasian continents collided, closing off the sea entirely beginning about 17 million years ago. Thick sediments accumulating in the Tethys Sea between Gondwana and Laurasia buckled and uplifted into mountain belts on the northern and southern flanks as the continents approached each other. The contact between the continents spurred a major mountain-building episode that raised the Alps and other ranges in Europe and squeezed out the Tethys Sea.

When the Tethys linking the Indian and Atlantic Oceans closed as Africa rammed into Eurasia, the collision resulted in the development of two major inland seas. These were the ancestral Mediterranean and a composite of the Black, Caspian, and Aral Seas, called the Paratethys, which covered much of eastern Europe. About 15 million years ago, the Mediterranean separated from the Paratethys, which became a brackish (slightly salty) sea, much like the Black Sea of today. About 6 million years ago, the Mediterranean Basin was completely cut off from the Atlantic Ocean when an isthmus created at Gibraltar by the northward movement of the African plate formed a dam across the strait. Nearly 1 million cubic miles of seawater evaporated, almost completely emptying the basin over a period of about 1,000 years.

The adjacent Black Sea might have had a similar fate. Like the Mediterranean, it is a remnant of an ancient equatorial body of water that separated Africa from Europe. The waters of the Black Sea drained into the desiccated basin of the Mediterranean. In a brief moment in geologic time, the Black Sea practically became a dry basin. Then during the last ice age, it refilled again and became a freshwater lake. The brackish, largely stagnant sea occupying the basin today has evolved since the end of the last ice age.

THE ATLANTIC

Some 170 million years ago, a great rift developed in the present Caribbean region and began to separate Pangaea into today's continents (Fig. 20). The

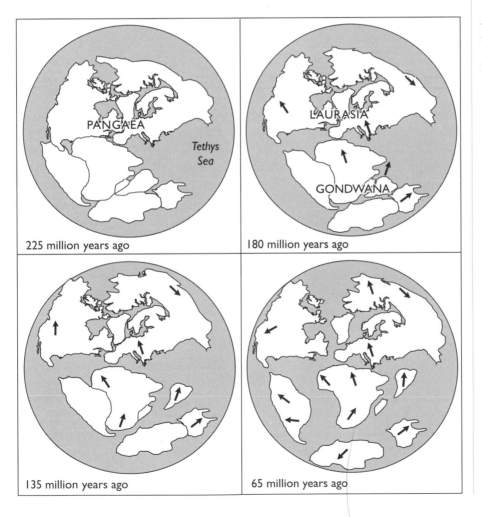

Figure 20 *The breakup of Pangaea 225, 180, 135, and 65 million years ago.*

225 million years ago

180 million years ago

135 million years ago

65 million years ago

27

breakup of Pangaea compressed the ocean basins, causing a rise in sea levels and a transgression of the seas onto the land. After the breakup, rather than separating at a constant speed, the continents drifted apart in spurts. The rate of seafloor spreading in the Atlantic matches the rate of plate subduction in the Pacific, where one plate dives under another, forming a deep trench. Following the breakup of Pangaea in the early Jurassic period about 170 million years ago, the Pacific plate was hardly larger than the present-day United States. The rest of the ocean floor was composed of other unknown plates that disappeared as the Pacific plate grew. The subduction of old oceanic crust explains why the ocean floor is no older than Jurassic in age.

The rift sliced northward through the continental crust that connected North America, northwest Africa, and Eurasia during the separation of the continents. In the process, this area breached and flooded with seawater, forming the infant North Atlantic. The rifting occurred over a period of several million years along a zone hundreds of miles wide. At about the same time, India, nestled between Africa and Antarctica, drifted away from Gondwana. While still attached to Australia, Antarctica swung away from Africa toward the southeast, forming the proto–Indian Ocean.

About 50 million years after rifting began, the infant North Atlantic had achieved a depth of 2 miles or more. It was bisected by an active midocean ridge system that produced new oceanic crust as the plates carrying the surrounding continents separated. Meanwhile, the South Atlantic began to form, opening up like a zipper from south to north. The rift propagated northward at a rate of several inches per year, similar to the separation rate of the two plates carrying South America away from Africa. The entire process of opening the South Atlantic took place in a span of just 5 million years.

The South Atlantic continued to widen as more than 1,500 miles of ocean separated South America and Africa. Africa moved northward, leaving Antarctica (still joined to Australia) behind, and began to close the Tethys Sea. In the early Tertiary, Antarctica and Australia broke away from South America and moved eastward. Afterward, the two continents rifted apart, with Antarctica moving toward the South Pole, while Australia continued moving northeastward.

By 80 million years ago, the North Atlantic was a fully developed ocean. Some 20 million years later, the Mid-Atlantic Rift progressed into the Arctic Basin. It detached Greenland from Europe, resulting in extensive volcanic activity (Fig. 21). North America was no longer connected with Europe except for a land bridge across Greenland that enabled the migration of species between the two continents. The separation of Greenland from Europe might have drained frigid Arctic waters into the North Atlantic, significantly lowering its temperature.

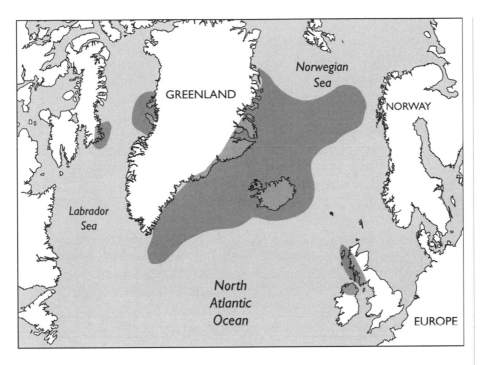

Figure 21 *Extensive volcanic activity during the opening of the North Atlantic 57 million years ago.*

The climate grew much colder. The seas withdrew from the land as the ocean dropped about 1,000 feet to perhaps its lowest level since the last several hundred million years and remained depressed for the next 5 million years. The drop in sea level also coincided with the accumulation of massive ice sheets atop Antarctica when it drifted over the South Pole. Meanwhile, the strait between Alaska and Asia narrowed, creating the nearly landlocked Arctic Ocean.

When Antarctica separated from South America and Australia and drifted over the South Pole some 40 million years ago, the polar vortex formed a circumpolar Antarctic ocean current. This current isolated the frozen continent, preventing it from receiving warm poleward flowing waters from the tropics. Since it was deprived of warmth, Antarctica became a frozen wasteland (Fig. 22). During this time, warm saltwater filled the ocean depths while cooler water covered the upper layers.

The Red Sea began to separate Arabia from Africa 34 million years ago, rapidly opening up from south to north. Prior to the opening of the Red Sea and Gulf of Aden, massive floods of basalt covered some 300,000 square miles of Ethiopia, beginning about 35 million years ago. The East African Rift Valley extending from the shores of Mozambique to the Red Sea split to form the Afar Triangle in Ethiopia. For the past 25 to 30 million years, Afar has been stewing with volcanism. An expanding mass of molten magma lying just beneath the crust uplifted much of the area thousands of feet.

Figure 22 *A view westward over Daniell Peninsula, Antarctica.*

(Photo by W. B. Hamilton, courtesy USGS)

Greenland was largely ice free until about 8 million years ago. At that time, a sheet of ice began building up to 2 miles thick and buried the world's largest island. Alaska connected with eastern Siberia and closed off the Arctic basin from warm-water currents originating from the tropics, resulting in the formation of pack ice in the Arctic Ocean.

About 4 million years ago, the Panama Isthmus separating North and South America uplifted as oceanic plates collided. The barrier created by the land bridge isolated Atlantic and Pacific species. Extinctions impoverished the once rich faunas of the western Atlantic. The new landform halted the flow of cold-water currents from the Atlantic into the Pacific. This effect, along with the closing of the Arctic Ocean from warm Pacific currents, might have initiated the Pleistocene ice ages, when massive glaciers swept out of the polar regions and buried the northern lands.

After discussing the origin of Earth and the ocean along with the evolution of the different seas through geologic history, the next chapter follows the exploration of the ocean and the discoveries made on the seabed.

2

MARINE EXPLORATION
DISCOVERIES ON THE SEABED

This chapter examines major discoveries made on the floor of the ocean. Early geologists thought the ocean floor was a barren desert covered by thick, muddy sediments washed off the land and by debris of dead marine organisms raining down from above. After billions of years, the sediments were assumed to have accumulated into layers several miles thick. The deep waters of the ocean were believed to be a vast featureless plain, unbroken by ridges or valleys and interspersed by a few scattered volcanic islands.

As remote sensing technology improved, the view of the seabed grew much more accurate and complex, revealing midocean ridges grander than terrestrial mountain ranges and chasms deeper than any canyon on the land. The midocean ridges, with highly active volcanic activity, appeared to generate new oceanic crust. The deep-sea trenches, with extensive earthquake activity, seemed to devour old oceanic crust. Strange sea creatures were found on the deep seafloor, where previously no life was thought possible. Indeed, the bottom of the ocean was much more complicated than ever imagined.

EXPLORING THE OCEAN FLOOR

The Renaissance period of the 14th century in western Europe began a renewed inquiry into scientific phenomena and extensive maritime exploration. It culminated with the discovery of the New World and many uncharted realms. The ice-covered continent Antarctica was discovered more than two centuries ago. It was stumbled upon purely by accident, even though Greek scholars predicted its existence more than 2,000 years earlier. The British navigator James Cook discovered "terra incognita," or unknown land, in 1774, although heavy pack ice forced him to turn back before actually seeing the frozen continent. By the 1820s, sealers were hunting seals prized for their oil and pelts in the frigid waters around Antarctica.

The United States, Great Britain, France, and Russia sent exploratory expeditions that made the first official sightings of Antarctica. The Scottish explorer Sir James Clark Ross, who in 1839 attempted to find the South Magnetic Pole, commanded one of these expeditions. He drove his ships through 100 miles of pack ice on the Pacific side of the continent until finally emerging into open water known today as the Ross Sea in his honor. After finding his way blocked by an immense wall of ice 200 feet high and 250 miles long, Ross gave up his quest to the South Magnetic Pole, which unbeknownst to him lay some 300 miles inland from his position.

To navigate the oceans in the past, ships relied on wind and sails (Fig. 23). Benjamin Franklin made a quite remarkable discovery when he worked for the London post office prior to the American Revolutionary War. British mail packets sailing to New England took two weeks longer to make the journey than did American merchant ships. The American ships apparently discovered a faster route. American whalers first noticed a strange behavior in whales, which kept to the edges of what appeared to be an invisible stream in the ocean and did not attempt to cross it or swim against its current.

Meanwhile, British captains, unaware of this stream, sailed in the middle of it. Sometimes, if the winds were weak, the ships were actually carried backward. The current was found to travel 13,000 miles clockwise around the North Atlantic basin at a speed of about 3 miles per hour. In 1769, Franklin had the current mapped, thinking it would be a valuable aid to shipping. After considering the crude methods of chart making in his days, Franklin's map of the Gulf Stream was unusually accurate. However, another century passed before any serious investigations of the current were ever conducted.

In the mid-1800s, depth soundings of the ocean floor were taken in preparation for laying the first transcontinental telegraph cable linking the United States with Europe. The depth recordings indicated hills, valleys, and a middle Atlantic rise named Telegraph Plateau, where the ocean was supposed

Figure 23 *The Polish full-rigged ship* Dar Pomorza *underway in the Boston harbor.*

(Photo by M. Putnam, courtesy U.S. Navy)

to be the deepest. Sometimes, sections of the telegraph cable became buried under submarine slides and had to be brought to the surface for repair.

In 1874, the British cable-laying ship H.M.S. *Faraday* was attempting to mend a broken telegraph cable in the North Atlantic. The cable rested on the ocean floor at a depth of 2.5 miles, where it passed over a large rise, which was later named the Mid-Atlantic Ridge (Fig. 24). While grappling for the cable, the claws of the grapnel snagged on a rock. When the grapnel was finally freed and brought to the surface, clutched in one of its claws was a large chunk of black basalt, a common volcanic rock. This was an astonishing discovery because volcanoes were not supposed to be in this region of the Atlantic Ocean.

The British corvette H.M.S. *Challenger,* the first fully equipped oceanographic research vessel, was commissioned in 1872 to explore the world's oceans. The crew took depth soundings, using a hemp rope with a lead weight

Figure 24 The mid-Atlantic spreading-ridge system separated the New World from the Old World.

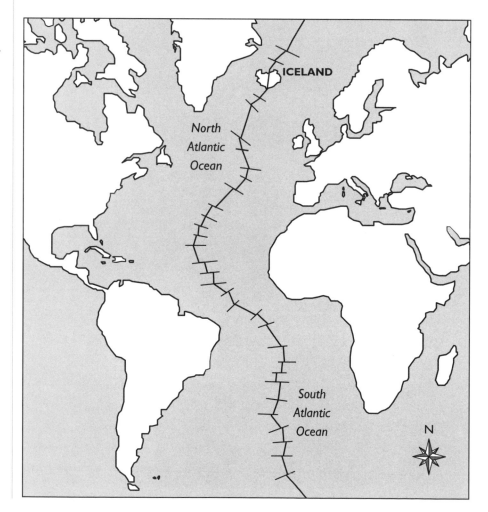

tied to one end and lowered over the side. They also took water samples and temperature readings. Additionally, they dredged bottom sediments for evidence of animal life living on the deep seafloor. The *Challenger's* nets hauled up a large number of deep-sea and bottom-dwelling animals, many from the deepest trenches. The catch included some of the strangest creatures, some of which were unknown to science or thought to have long gone extinct.

During nearly 4 years of exploration, the *Challenger* charted 140 square miles of ocean bottom and sounded every ocean except the Arctic. The deepest sounding was taken off the Mariana Islands in the western Pacific. While recovering samples in the deep waters off the Marianas, the research vessel encountered a deep trough known as the Mariana Trench, which forms a long line northward from the Island of Guam. It is the lowest place on Earth, reaching a depth of nearly 7 miles below sea level.

While dredging the deep ocean bottom in the Pacific, the *Challenger* recovered rocks resembling dense lumps of coal. After being mistaken for fossils or meteorites, the rocks were put on display in the British Museum as geologic oddities from the ocean floor. Almost a century later, further analysis showed the true value of the dark, potato-sized clumps. The nodules contained large quantities of valuable metals, including manganese, copper, nickel, cobalt, and zinc. Scientists realized that the world's largest reserve of manganese nodules lay on the bottom of the North Pacific, about 16,000 feet below the surface. Fields thousands of miles long contained nodules estimated at 10 billion tons.

Other valuable minerals were found on the deep-sea floor. In 1978, the French research submersible *Cyana* discovered unusual lava formations and mineral deposits on the seabed in the eastern Pacific more than 1.5 miles deep. These deposits were sulfide ores in 30-foot-high mounds of porous gray and brown material. The massive sulfide deposits contained abundant iron, copper, and zinc. The French research vessel *Sonne* found another sulfide ore field nearly 2,000 miles long on the floor of the East Pacific. The sediments contained as much as 40 percent zinc along with deposits of other metals, some in greater concentrations than their land-based counterparts.

Research vessels discovered valuable sediments more than 7,000 feet deep on the bed of the Red Sea (Fig. 25) between Sudan and Saudi Arabia. The largest deposit was in an area 3.5 miles wide known as the Atlantis II Deep, named for the research vessel that discovered it. The rich bottom ooze was estimated to contain about 2 million tons of zinc, 400,000 tons of copper, 9,000 tons of silver, and 80 tons of gold. The sea undoubtedly provides unheard-of mineral riches.

Much of the evidence for continental drift was found on the ocean floor. However, many early 20th-century geologists refuted the theory of continental drift. They believed that narrow land bridges spanned the distances between continents. Geologists used the similarity of fossils in South America

Figure 25 *The Red
Sea and the Gulf of Aden
are prototype seas created
by seafloor spreading.*

(Photo courtesy USGS
Earthquake Information
Bulletin)

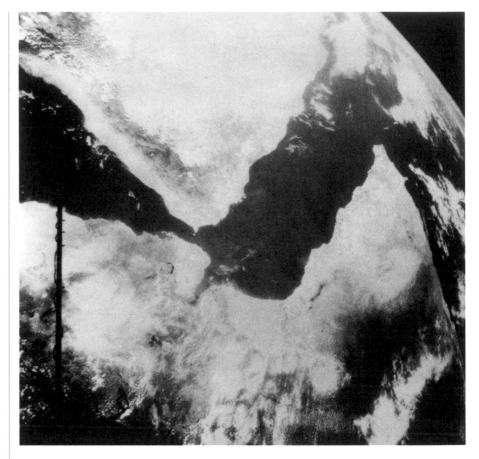

and Africa to support the existence of a land bridge between the two continents. The idea was that the continents were always fixed and that land bridges
rose from the ocean floor to enable species to migrate from one continent to
another. Later, the land bridges sank beneath the surface of the sea. However,
a search for evidence of land bridges by sampling the ocean floor failed to turn
up even a trace of sunken land.

   The German meteorologist and Arctic explorer Alfred Wegener argued
that a land bridge was not possible because the continents stand higher than
the seafloor for the simple reason that they are composed of light granitic
rocks that float on the denser basaltic rocks of the upper mantle. In 1908, the
American geologist Frank Taylor described an undersea mountain range
between South America and Africa, which became known as the Mid-
Atlantic Ridge. He believed it was a line of rifting between the two continents. The ridge remained stationary, while the two continents slowly crept
away from it in opposite directions.

Eventually, advances in technology allowed marine scientists to begin exploring the oceans firsthand. In 1930, the American naturalist and explorer William Beebe invented the first bathysphere. It held one person and could descend more than 3,000 feet, an unheard-of depth in those days. This crude submersible enabled scientists to observe strange new marine life. However, because it was tethered to a ship, its maneuverability was limited. Later, the U.S. Navy led efforts to develop deep-submergence vehicles that could operate on their own, which enhanced marine exploration considerably. In the 1960s, recognition of the value to science of piloted, free-ranging minisubmarines led to the birth of *Alvin* (Fig. 26), the

Figure 26 *The deep submersible* Alvin *at its Wood's Hole, Massachusetts, port.*

(Photo by R. A. Wahl, courtesy U.S. Navy)

workhorse for deep ocean exploration. The 23-foot-long submersible held three people, could descend some 2 miles deep, and could stay submerged for eight hours.

Even by the early 1970s, knowledge of the seafloor and the capacity to explore it were still rudimentary. Shipboard sonar was inadequate for mapping the rugged topography of the midocean ridges. The imagery improved substantially when sonar devices were mounted on a vehicle and towed behind a ship at a considerable depth. A system called SeaBeam made high-resolution sonar maps of the midocean ridge crests. Its sonar covered a broad swath of seafloor, allowing a ship to map an entire area by tracking back and forth in well-spaced lines.

Cameras were also mounted on undersea sleds (Fig. 27) and pulled through elaborate obstacle courses in the dark abyss. However, the instruments were damaged or lost at an alarming rate. A massive camera vehicle called *Angus* weighed 1.5 tons, enabling it to be towed almost directly beneath the ship for better navigational control. The most sophisticated device, called Deep Tow, carried sonar, television cameras, and sensors for measuring temperature, pressure, and electricity. During operation over the East Pacific Rise off the coast of Ecuador, the camera sled "flew" into a hot plume of water. Upon further exploration, photographs taken by *Angus* revealed a lava field scattered with large white clams.

When the submersible *Alvin* was sent down to investigate this phenomenon, it discovered an oasis of hydrothermal vents (Fig. 28) and exotic deep-sea creatures 1.5 miles below sea level. The base of jagged basalt cliffs showed evidence of active lava flows, including fields strewn with pillow lavas. Unusual chimneys called black smokers spewed out hot water blackened with sulfide minerals. Others called white smokers ejected milky, white-hot water. Species previously unknown to science lived in total darkness among the hydrothermal vents. Tube worms growing up to 10 feet tall swayed in the hydrothermal currents. Giant crabs scampered blindly across the volcanic terrain. Huge clams growing up to 1 foot long and clusters of mussels formed large communities around the vents.

In other areas of the ocean, scientists made other remarkable discoveries. Biologists of the Smithsonian Institute using a deep-sea submersible made a surprising discovery in 1983 off the Bahamas. A totally new and unexpected form of algae lived on an uncharted seamount at a depth of about 900 feet, deeper than any previously known marine plant larger than a microbe. The species comprised a variety of purple algae with a unique structure. It consisted of heavily calcified lateral walls and very thin upper and lower walls. The cells grew on top of each other, similar to cans stacked at a grocery store, for maximum surface exposure to the feeble sunlight. The discovery expanded the

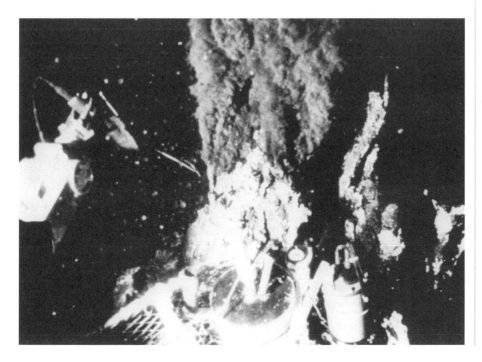

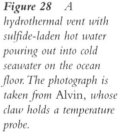

Figure 28   *A hydrothermal vent with sulfide-laden hot water pouring out into cold seawater on the ocean floor. The photograph is taken from* Alvin, *whose claw holds a temperature probe.*

(Photo by N. P. Edgar, courtesy USGS)

role that algae play in the productivity of the oceans, marine food chains, sedimentary processes, and reef building.

The 274-foot-long research vessel *Atlantis* was the first ship of its kind to support both manned submersibles, such as the renowned *Alvin,* and unmanned remotely operated undersea vehicles. Operated by the Woods Hole Oceanographic Institute, *Atlantis* combined technologies that used to be operated on separate vessels. Instead of surveying a site with one research tool and returning months later with another, *Atlantis* could conduct many kinds of research during a single visit. For instance, during a survey of the ocean floor, the ship would tow cameras, then use a remotely operated vehicle for the initial survey, followed up by manned submersibles to take larger samples, thus saving time and expense.

SURVEYING THE SEABED

The more scientists probed the ocean floor, the more complex it turned out to be. The ocean covers about 70 percent of Earth's surface to an average depth of over 2 miles. It is shallowest in the Atlantic basin and deepest in the Pacific basin. If Mount Everest, the world's tallest mountain, were placed into the deepest part of the Pacific basin, the water would still rise about 1 mile above its peak. Yet in relation to the overall size of Earth, the ocean is merely a thin veneer of water comparable to the outer skin of an onion.

Early methods of sampling the seabed included dragging a dredge behind a ship to scoop up the bottom sediments or using a device called a snapper (Fig. 29), whose jaws automatically closed when the instrument struck the bottom. However, these techniques sampled only the topmost layers, which could not be recovered in the order of their original deposition. In the early 1940s, Swedish scientists developed a piston corer. When dropped to the seabed, it retrieved a vertical section of the ocean floor intact. The corer consisted of a long barrow that plunged into the bottom mud under its own weight. A piston firing upward from the lower end of the barrow sucked up sediments into a pipe, and the core samples were then brought to the surface (Fig. 30 and Fig. 31).

The bottom of the ocean was first thought to contain sediments several miles thick that washed off the continents after billions of years of accumulation. However, core drilling at several sites revealed that the oldest sediments were less than 200 million years old. The sediments were measured with an undersea device that used seismic waves similar to sound waves to locate sedimentary structures. An ocean bottom seismograph (Fig. 32) dropped to the seafloor recorded microearthquakes in Earth's submarine crust and rose automatically to the surface for recovery. Seismic instruments towed behind ships

Figure 29 *A snapper sampling instrument, whose jaws close when striking the bottom.*

(Photo by K. O. Emery, courtesy USGS)

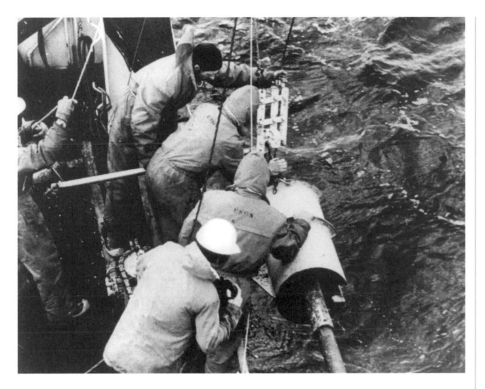

Figure 30 *Piston coring in the Gulf of Alaska.*

(Photo by P. R. Carlson, courtesy USGS)

also detected geologic structures deep within the suboceanic crust. These surveys provided important information about the ocean floor that could not be obtained by direct means. They revealed that instead of miles of silt and mud, the oceanic crust contained only a few thousand feet of sediment.

During the height of the cold war in the late 1950s, American and Russian oceanographic vessels mapped the ocean floor to enable ballistic missile submarines to navigate in deep water without grounding on uncharted seamounts. During heightened cold war tensions, when Russian aircraft shot down a Korean civilian airliner over Sakhalin Island on August 30, 1983, killing all 269 passengers and crew, an unprecedented search for the downed aircraft was conducted using the robotic submersible *Deep Drone* (Fig. 33) operated by the U.S. Navy.

Sonar depth ranging was another important tool for mapping undersea terrain. SeaMarc, a side-looking sonar system towed in a "fish" about 1,000 feet above the ocean floor, provided a sonar image of the ocean bottom (Fig. 34) by bouncing sound waves off the seabed. As ships traversed the Atlantic Ocean, onboard sonographs painted a remarkable picture of the ocean floor. Lying 2.5 miles deep in the middle of the Atlantic Ocean was a huge submarine mountain range, surpassing in scale the Alps and the Himalayas com-

Figure 31   *A piston corer on the ocean floor.*

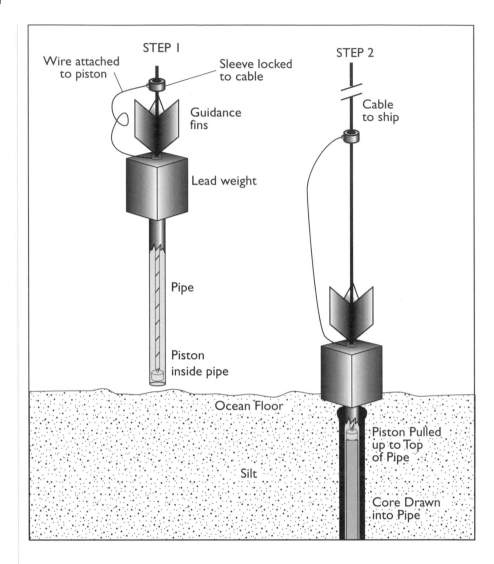

bined. The range ran down the middle of the ocean floor, weaving halfway between the continents that surrounded the Atlantic basin. This massive ridge was discovered to be the site of intense volcanic activity, as though Earth's insides were coming out.

The midocean ridges were found to be a string of seamounts in a region where scientists had assumed that the deep seafloor should have been flat and barren. With more detailed mapping of the ocean floor, scientists found that the Mid-Atlantic Ridge was the most peculiar mountain range yet discovered. The ridge crest was 10,000 feet above the ocean floor. A deep trough ran through the middle of it like a giant crack in Earth's crust. The crevasse was 4 miles deep in places, or four times deeper than the Grand Canyon, and up to 15 miles wide, making it the grandest canyon on Earth.

Undersea surveys have shown that the submerged mountains and undersea ridges formed a continuous chain 46,000 miles long, several hundred miles wide, and up to 10,000 feet high that winds around the globe like the stitching on a baseball. Although the midocean ridge system lies deep beneath the sea, it is easily the most dominant feature on the face of the planet. It extended

Figure 32 *An ocean bottom seismograph provides direct observations of earthquakes on midocean ridges.*

(Photo courtesy USGS)

Figure 33   *The submersible* Deep Drone *being launched to explore for Korean Air Lines Flight 007 shot down near Sakhalin Island on August 30, 1983, by Russian aircraft.*

(Photo by F. Barbante, courtesy U.S. Navy)

over an area greater than that covered by all major terrestrial mountain ranges combined. Moreover, it exhibited many unusual features, including massive peaks, sawtooth ridges, earthquake-fractured cliffs, deep valleys, and a variety of lava formations. Along much of its length, the ridge system is carved down the middle by a sharp break or rift that is the center of an intense heat flow. Furthermore, the midocean ridges were the sites of frequent earthquakes and volcanic eruptions, as though the entire system was a series of giant cracks in Earth's crust, where lava oozed out onto the ocean floor.

When advanced instrumentation was developed, the view of the seafloor came into better focus. The ocean floor proved to be far more active and younger than previously imagined. Additional surveys conducted across the extensive undersea mountain range included rock sampling, sonar depth finding, thermal measurements, magnetic readings, and seismic surveys. The resulting data suggested that the oceanic crust was spreading outward at the midocean ridge. Magma rising from the mantle erupted onto the ocean floor, adding new oceanic crust to the ridge crest as both sides pulled apart.

Temperature surveys showed anomalous amounts of heat seeping out of Earth in the mountainous regions of the middle Atlantic, as though magma bled from the mantle through cracks in the oceanic crust. Volcanic activity in the ridges suggested that new material was being added to the seafloor. This activity appeared to be more intense in the Atlantic Ocean, where the midocean

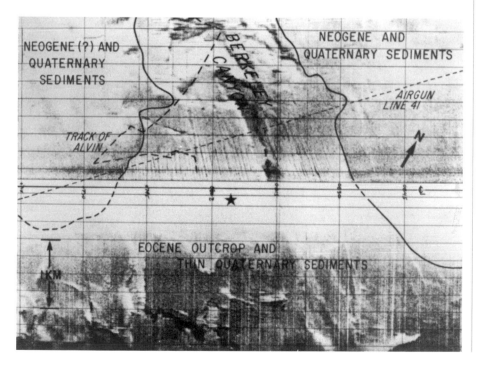

Figure 34 Sonograph of the lower continental slope off the Atlantic coast from SeaMarc.

(Photo by N. P. Edgar, courtesy USGS)

ridge is steeper and more jagged, than in the Pacific or the Indian Oceans, where branches of oceanic ridges were overridden by continents.

Deep-sea trenches off continental margins and volcanic island arcs were originally thought to have been created by the tremendous weight of sediments washed off the continents and pulled down into the mantle by a dense, underlying material. The downward pull on the sediments formed vast bulges in the ocean floor called geosynclines. However, gravity surveys conducted over the trenches indicated that the pull of gravity was much too weak to account for the sagging of the seafloor.

The trenches were also found to be sites of almost continuous earthquake activity deep in the bowels of Earth. The deep-seated earthquakes acted like beacons marking the boundaries of a large slab of crust descending into the mantle. The unusual activity of the trenches suggested that they were sites where old oceanic crust subducted into Earth's interior. Perhaps here at last was the engine that drove the continents around the surface of Earth.

GEOLOGIC OBSERVATIONS

Observations of these and other fascinating geologic features on the ocean floor led to the development of the seafloor-spreading theory. The hypothesis described the creation and destruction of the ocean floor at specific regions around the world. The seafloor-spreading theory resolved many problems connected with the mysterious characteristics on the seafloor. These include the midocean ridges, the relatively young ages of rocks in the oceanic crust, and the formation of island arcs. However, more importantly, here at last was the long-sought mechanism for continental drift (Table 5). The continents do not plow through the ocean crust like icebreakers slicing through frozen seas, as previously thought. Instead, they ride above a pliable mantle like ships caught in mobile ice floes.

Exploration of the ocean floor ushered in a new understanding of the forces that shaped the planet. After overwhelming geologic and geophysical evidence was collected from the floor of the ocean in support of the theory of continental drift, geologists finally abandoned the archaic thinking of the 19th century. By the late 1960s, most geologists in the Northern Hemisphere, who had long rejected the theory, finally joined their southern colleagues, who were long convinced of the reality of continental drift because of overwhelming evidence in South America and the opposing African continent.

The discovery of many mysteries on the seabed, including spreading ridges and deep-sea trenches, led geologists to develop an entirely new way of looking at Earth, called the theory of plate tectonics (Fig. 35). Tectonics, from the Greek word *tekton,* meaning "builder," is the geologic process responsible

TABLE 5   CONTINENTAL DRIFT

	Geologic Division (millions of years)	Gondwana	Laurasia
Quaternary	5		Opening of the Gulf of California
Pliocene	11	Spreading begins near the Galápagos Islands	Spreading changes directions in the eastern Pacific
		Opening of the Gulf of Aden	Birth of Iceland
Miocene	26		
		Opening of the Red Sea	
Oligocene	37		
		Collision of India with Eurasia	Spreading begins in the Arctic Basin
Eocene	54		Separation of Greenland from Norway
Paleocene	65	Separation of Australia from Antarctica	
			Opening of the Labrador Sea
		Separation of New Zealand from Antarctica	
			Opening of the Bay of Biscay
		Separation of Africa from Madagascar and South America	Major rifting of North America from Eurasia
Cretaceous	135		
		Separation of Africa from India, Australia, New Zealand, and Antarctica	Separation of North America from Africa begins
Jurassic	180		
Triassic	250	Assembly of all continents into the supercontinent Pangaea	

for features on Earth's surface. The theory incorporated the process of seafloor spreading and continental drift into a comprehensive model. Therefore, all aspects of Earth's history and structure could be unified by the revolutionary concept of movable plates.

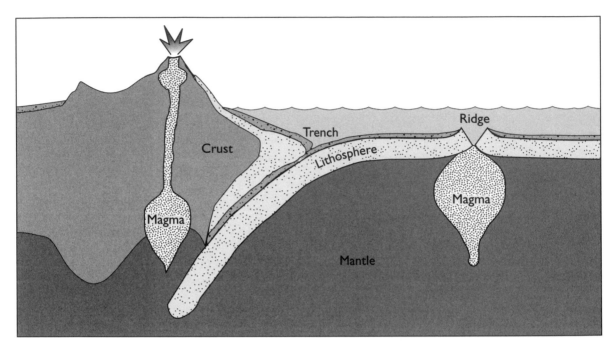

Figure 35 *The plate tectonics model. New oceanic crust is generated at spreading ridges, and old oceanic crust is destroyed in subduction zones, which moves the continents around Earth.*

The Atlantic Ocean is bisected by the Mid-Atlantic Ridge, which manufactures new oceanic crust as the continents surrounding the Atlantic basin spread apart. The Mid-Atlantic Ridge is the center of intense seismic and volcanic activity. It is the focus of high heat flow from Earth's interior. Molten magma originating from the mantle rises through the lithosphere and erupts onto the ocean floor, adding new oceanic crust to both sides of the ridge crest.

As the Atlantic basin widens, the surrounding continents separate at a rate of about 1 inch per year. In response to the widening seafloor in the Atlantic and the separation of continents around the Atlantic basin, the Pacific basin shrinks at a corresponding rate. Subduction zones (Fig. 36) that destroy old oceanic crust in deep-sea trenches ring the Pacific. Spreading ridges in the Pacific are also much more active than those in the Atlantic. These features on the ocean floor are responsible for most of the geologic activity that surrounds the Pacific Ocean.

OCEAN DRILLING

The oceans have an average depth of more than 2 miles and are blanketed by layers of thick sediments. To date these sediments correctly, they must be recovered in the order they were originally laid down, thus dredging techniques were of little use. Fortunately, a system known as seafloor coring was developed, enabling scientists to take accurate sediment samples. A hollow pipe is drilled into the sediments, and a long cylindrical sample is brought to the surface. Early attempts at coring in deep water, however, penetrated only a few feet into the upper sediments of the ocean floor.

In the mid-1960s, the National Science Foundation sponsored a deep-sea drilling program called Project Mohole. The Mohorovicic discontinuity, or simply Moho (Fig. 37), named for the Yugoslav seismologist Andrija Mohorovicic, is the point of contact between Earth's crust and mantle. The crust is the thinnest in the oceans, measuring only about 3 to 5 miles thick. Scientists hoped that the Moho would provide new clues about the origin, age, and composition of Earth's interior, which land-based drilling could not obtain. Unfortunately, the task of drilling through miles of oceanic crust in waters as much as 2 miles deep or more became expensive and time-consuming.

In 1968, the British research vessel *Glomar Challenger* was commissioned for the Deep Sea Drilling Project, a consortium of American oceanographic

Figure 36 *The subduction of the ocean floor provides new molten magma for volcanoes that fringe the deep-sea trenches.*

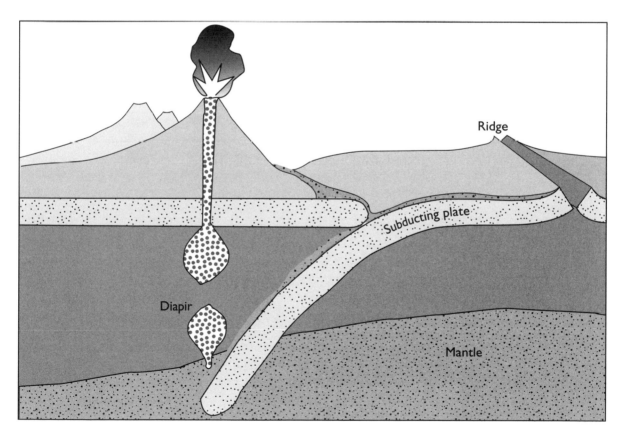

49

Figure 37 *The deep structure of Earth's crust, showing the position of the Moho.*

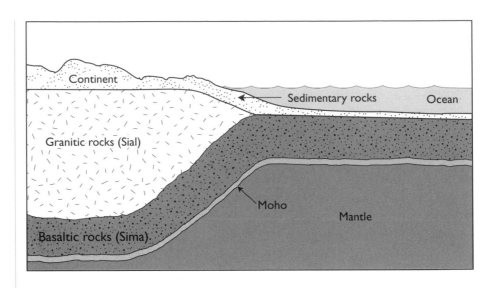

institutions. The project's objective was to drill a large number of shallow holes in widely scattered parts of the ocean floor in an attempt to prove the theory of seafloor spreading. A similar deep-sea drillship called the *Glomar Pacific* (Fig. 38) was the first to begin drilling on the Atlantic outer continental shelf and slope of the United States. Both ships were designed with a 140-foot drilling derrick amidships. Computerized thrusters located fore and aft maintained station over the drill hole even in rough seas.

A string of drill pipe dangled as much as 4 miles beneath the ship, with the drill bit cutting through the sediment by the force of its own weight. The core, which is a cylindrical vertical section of rock, was retrieved through the drill stem by a removable inner barrel, allowing the drill bit to remain in the hole. When the drill bit became dulled, it and the drill pipe were brought back up to the surface for replacement. The drill string was then lowered back over the drill hole, and a special funnel-like apparatus guided the drill bit into the hole.

The primary purpose of the international Ocean Drilling Program (ODP) and the Joint Oceanographic Investigation for Deep Earth Sampling (JOIDES) was to take rotary core samples of the ocean floor at hundreds of sites around the world. However, precautions had to be taken not to drill in potentially productive oil fields, where drilling might cause blowouts that would result in hazardous oil spills.

Just the opposite situation occurred on the south flank of the Costa Rica Rift east of the Galápagos Islands in 1979 when the *Challenger* drilled a hole into the crust. Instead of blowing out hot water, which is often the case, the well sucked in a powerful, steady stream of seawater. The suction resulted from the downward convection by the circulating water within the oceanic crust as it descended toward a magma chamber, where it acquired heat during hydrothermal activity.

The deepest hole was bored into the ocean floor in the eastern Pacific near the Galápagos Islands by the drillship JOIDES *Resolution*. Its purpose was to sample a section of the entire oceanic crust from top to bottom in an area where it was though to be the thinnest. During a 14-year period beginning in 1979, the ship made seven trips to the drill site to deepen the hole, with each session lasting up to two months. On the sixth trip, the ship first had to recover drill pipe lost in the hole during the previous effort. When this task was accomplished, the hole was further extended to a depth of more than 6,500 feet beneath the seabed. In January 1993, the *Resolution* returned again to deepen the hole another 370 feet only to lose the drill bit. This mishap forced the crew to abandon the drill hole perhaps only a few hundred feet short of their goal.

Figure 38 *The* Glomar Pacific *drilling on the Atlantic outer continental shelf and slope of the United States.*

(Photo courtesy USGS)

While taking a shortcut to the bottom of the ocean crust, ODP scientists found a site where the lower crust is uncovered along the Atlantis II Fracture Zone in the Indian Ocean. This is part of the midocean ridge that forms the boundary between the African and Antarctic tectonic plates. Running down the middle of the ridge is a feature called a spreading center, which periodically breaks apart, leaving a gap that fills with molten magma. As the magma cools and hardens, the rock forms new oceanic crust that joins to the ends of the plates.

The structure of the spreading center resembles steps in a staircase, with short, straight segments roughly parallel to each other (see Fig. 72). The fracture zones are valleys that connect adjacent segments like the vertical jumps between the steps. When the scientists drilled through the valley floor of the fracture zone, they recovered coarsely crystalline rocks called gabbros. These are composed of iron-magnesium silicates known to make up the lower segment of the ocean crust.

After recovering and dating cores from several midocean ridges, the *Challenger* made a truly remarkable discovery. The farther away from the deep-sea ridges the ship drilled, the thicker and older the sediments became. Even more surprising was that the thickest and oldest sediments were not billions of years old as expected but were in fact younger than 200 million years. Near the continental shelves, where thick layers of sediments form flat abyssal plains, the drill cores revealed thin beds of calcium carbonate just above hard volcanic rock that was buried under thousands of feet of red clay and other sediments. The discovery of abyssal red clay, whose color signifies a terrestrial origin, provided additional evidence for seafloor spreading.

The deepest abysses in the world are next to continental margins. These are the actual boundaries of continents, where the oceanic lithosphere is the oldest. The calcium carbonate layer located by the *Challenger* was about 4 miles deep, far below the depth where the crush of cold water dissolves calcium carbonate. While well protected from the corrosive effect of seawater by the overlying sediments, the calcium carbonate originating in shallower water near midocean ridges was somehow transported to the edges of the continents.

The floor of the Atlantic conveys lithosphere, the rigid layer of the upper mantle, away from its point of origin at the Mid-Atlantic Ridge. The ocean floor at the crest of the midocean ridge consists mostly of basalt, a black volcanic rock. When continuing away from the crest, the bare rock is blanketed by an increasing thickness of sediments. These are composed mostly of red clay from detritus material washed off the continents and from windblown desert sediments landing in the sea. Some large sandstorms originating over the Sahara Desert blow dust so high into the atmosphere that prevailing air currents carry the dust completely across the Atlantic Ocean to South America

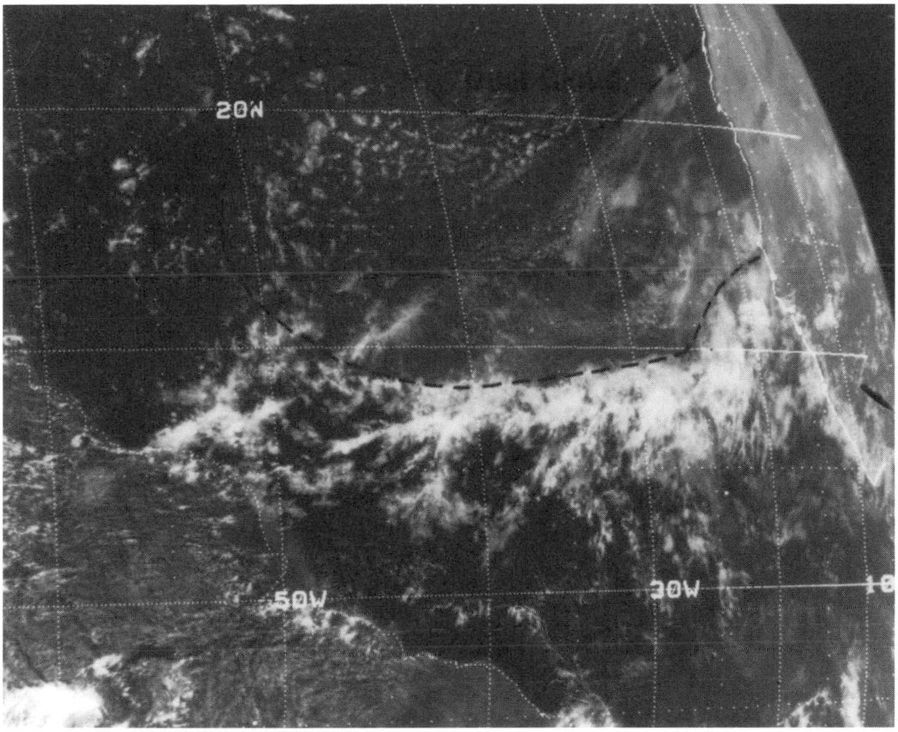

Figure 39 During the summer of 1976, drought conditions in West Africa and a prevailing easterly wind resulted in a dust surge—an enormous cloud of dust blowing out over the Atlantic Ocean from the Sahara Desert.

(Photo courtesy NOAA)

(Fig. 39), where about 13 million tons land in the Amazon basin annually. Fast-moving storm systems in the Amazon rain forest pull in the African dust, which contains nutrients that enrich the soil.

Near the ridge crest, the sediments are predominantly composed of calcareous ooze built up by a rain of decomposed shells and skeletons of microorganisms. Further away from the ridge crest, the slope falls below the calcium carbonate compensation zone generally about 3 miles deep. Below this depth, calcium carbonate, whose solubility increases with pressure, dissolves in seawater. Therefore, only red clay should exist in the deep abyssal waters far from the crest of the midocean ridge.

Yet drill cores taken from the abyssal plains near continental shelves, where the oceanic crust is the oldest and deepest, clearly showed thin layers of calcium carbonate below thick beds of red clay and above hard volcanic rock. Geologists concluded that the red clay protected the calcium carbonate from dissolving in the deep waters of the abyss. The discovery implies that the midocean ridge was the source of the calcium carbonate discovered near continental margins and that the seafloor moved across the Atlantic basin.

MAGNETIC SURVEYS

Geologists looking for a decisive test for seafloor spreading stumbled upon magnetic reversals on the ocean floor. Recognition of the reversal of the geomagnetic field began in the early 1950s. In 1963, the British geologists Fred Vine and Drummond Mathews thought that magnetic reversal would be a decisive test for seafloor spreading. Experiments using sensitive magnetic recording instruments called magnetometers towed behind ships over the midocean ridges (Fig. 40) revealed magnetic patterns locked in the volcanic rocks on the seafloor. These patterns alternated from north to south and were mirror images of each other on both sides of the ridge crest. The magnetic fields captured in the rocks also showed the past position of the magnetic poles as well as their polarities.

As the iron-rich basalts of the midocean ridges cool, the magnetic fields of their iron molecules line up in the direction of Earth's magnetic field at the time of their deposition. As the ocean floor spreads out on both sides of the ridge, the basalts solidify. They establish a record of the geomagnetic field at each successive reversal, somewhat like a magnetic tape recording of the his-

Figure 40 *A crew member lowers a magnetometer over the stern of the oceanographic research ship USNS Hayes.*

(Photo courtesy U.S. Navy)

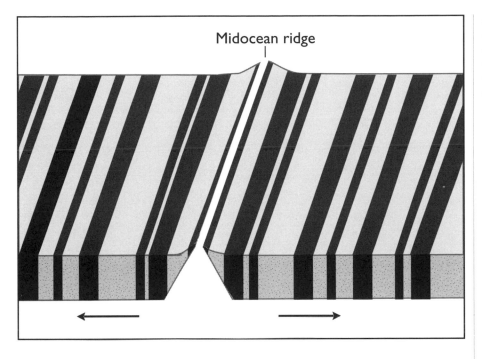

Midocean ridge

tory of the geomagnetic field. Normal polarities in the rocks are reinforced by the present magnetic field, while reversed polarities are weakened by it. This process produced parallel bands of magnetic rocks of varying width and magnitude on both sides of the ridge crest (Fig. 41). Here at last was clinching proof for seafloor spreading. In order for the magnetic stripes to form in such a manner, the ocean floor had to be pulling apart.

Two or three times every million years, Earth's geomagnetic field reverses polarity, with the north and south magnetic poles switching places. Over the last 4 million years, the field reversed 11 times. Over the last 170 million years, Earth's magnetic field has reversed 300 times. No reversals occurred during long stretches of the Permian and Cretaceous periods. Furthermore, a sudden polar shift of 10 to 15 degrees occurred between 100 million and 70 million years ago.

Since about 90 million years ago, reversals have steadily become more frequent, and the polar wandering has decreased to only about 5 degrees. The last time the geomagnetic field reversed was about 780,000 years ago, and Earth appears to be well overdo for another one. The magnetic field in existence 2,000 years ago was considerably stronger than it is today. Earth's magnetic field seems to have weakened over the past 150 years, amounting to a loss of about 1 percent per decade. If the present rate of decay continues, the field could reach zero and go into another reversal within the next 1,000 years or so.

TABLE 6   COMPARISON OF MAGNETIC REVERSALS WITH OTHER PHENOMENA (DATES IN MILLIONS OF YEARS)

Magnetic Reversal	Unusual Cold	Meteorite Activity	Sea Level Drops	Mass Extinctions
0.7	0.7	0.7		
1.9	1.9	1.9		
2.0	2.0			
10				11
40			37–20	37
70			70–60	65
130			132–125	137
160			165–140	173

The magnetic stripes also provided a means of dating practically the entire ocean floor. This is because the magnetic reversals occur randomly and any set of patterns is unique in geologic history (Table 6). The rate of seafloor spreading was calculated by determining the age of the magnetic stripes by dating drill cores taken from the midocean ridge and measuring the distance from their points of origin at the ridge crest. During the past 100 million years, the rate of seafloor spreading has changed little. Periods of increased acceleration were accompanied by an increase in volcanic activity. During the past 10 to 20 million years, a progressive acceleration has occurred, reaching a peak about 2 million years ago.

The spreading rates on the East-Pacific Rise are upward of 6 inches per year, which results in less topographical relief on the ocean floor. The active tectonic zone of a fast-spreading ridge is usually quite narrow, generally less that 4 miles wide. In the Atlantic, the rates are much slower, only about 1 inch per year. This allows taller ridges to form. Calculating the rate of seafloor spreading for the Atlantic indicates that it began to open around 170 million years ago—a time span remarkably concurrent with the estimated date for the breakup of the continents.

SATELLITE MAPPING

In 1978, the radar satellite *Seasat* (Fig. 42) precisely measured the distance to the ocean surface over most of the globe. Buried structures beneath the ocean floor appeared in full view for the first time. Among the astonishing discoveries was the fact that ridges and trenches on the ocean bottom produce corresponding hills and valleys on the surface of the ocean because of variations in the pull of gravity. The topography of the ocean surface showed bulges and depressions with a relief between highs and lows as much as 600 feet. How-

Figure 42 *Artist's concept of the* Seasat A *satellite as it studies the oceans from Earth orbit.*

(Photo courtesy USGS)

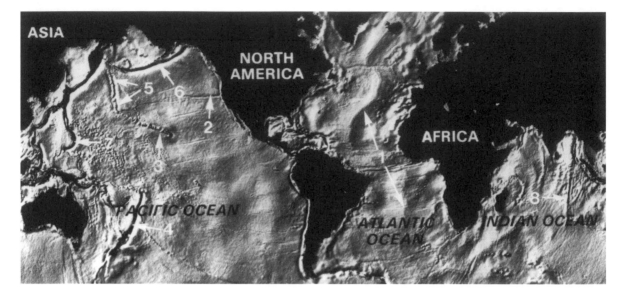

Figure 43 Radar altimeter data from the Geodynamic Experimental Ocean Satellite (GEOS-3) and Seasat *produced this map of the ocean floor. (1) Mid-Atlantic Ridge, (2) Mendocino Fracture Zone, (3) Hawaiian Islands, (4) Tonga Trench, (5) Emperor Seamounts, (6) Aleutian Trench, (7) Mariana Trench, (8) Ninety East Ridge.*

(Photo courtesy NASA)

ever, because these surface variations range over a wide area, they are generally unrecognized on the open sea.

The pull of gravity from undersea mountains, ridges, trenches, and other structures of varying mass distributed over the seafloor controls the shape of the surface water. Undersea mountain ranges produce large gravitational forces that cause seawater to pile up around them, resulting in gentle swells on the ocean surface. Conversely, submarine trenches with less mass to attract water form shallow troughs in the sea surface. For example, a trench 1 mile deep can cause the ocean to drop dozens of feet. A gravity low, a deviation of the gravity value from the theoretical value, formed as a plate sinks into the mantle off Somalia in northwest Africa might well be the oldest trench in the world.

The satellite altimetry data was used to produce a map of the entire ocean surface (Fig. 43), representing the seafloor as much as 7 miles deep. Chains of midocean ridges and deep-sea trenches were delineated with a clarity greater than had been achieved with any other method of mapping the ocean floor. The seafloor maps also uncovered many new features such as rifts, ridges, seamounts, and fracture zones and better defined several known features. The maps provided additional support for the theory of plate tectonics. This theory holds that the crust is broken into several plates whose constant shifting is responsible for the geologic activity on Earth's surface, including the growth of mountain ranges and the widening of ocean basins.

The satellite imagery also revealed long-buried parallel fracture zones undiscovered by conventional seafloor-mapping techniques. The faint lines running like a comb through the central Pacific seafloor might be controlled by convection currents in the mantle 30 to 90 miles beneath the oceanic crust. Each circulating loop consists of hot material rising and cooler material sinking back into the depths, tugging on the ocean floor as it descends.

The data also revealed a fracture zone in the southern Indian Ocean that shows India's break from Antarctica around 180 million years ago. The 1,000-mile-long gash, located southwest of the Kerguelen Islands, was gouged out of the ocean floor as the Indian subcontinent inched northward. When India collided with Asia, more than 100 million years after it was set adrift, it pushed up the Himalaya Mountains to great heights like squeezing an accordion. A strange series of east-west wrinkles in the ocean crust just south of India verifies that the Indian plate is still pushing northward, continuously raising the Himalayas and shrinking the Asian continent by as much as 3 inches a year.

Even buried structures came into full view for the first time. One example is an ancient midocean ridge that formed when South America, Africa, and Antarctica began separating around 125 million years ago. The seafloor-spreading center was buried deep under thick layers of sediments. The boundary between the plates moved westward, leaving behind the ancient ridge, which began to subside. The ridge's discovery might help trace the evolution of the oceans and continents over the last 200 million years. The satellite data provided further proof that the deep-sea floor remains, in large part, uncharted territory and that the exploration of inner space is just as important as the exploration of outer space.

After exploring the ocean floor, the next chapter searches the seabed for evidence for plate tectonics, the force that moves great chunks of crust around the surface of Earth and that is responsible for geologic activity on the planet.

3

THE DYNAMIC SEAFLOOR

THE OCEANIC CRUST

This chapter examines the ocean's role in plate tectonics, the process that changes the face of Earth. The ocean's crust is constantly changing. It is relatively young compared with continental crust and is less than 5 percent of Earth's age. The age difference is due to the recycling of oceanic crust into the mantle. Almost all the ocean floor has disappeared into Earth's interior over the last 170 million years. The oceanic crust is continuously being created at midocean ridges, where basalt oozes out of the mantle through rifts in the crust. It is destroyed in deep-sea trenches, where the lithosphere plunges into the mantle and remelts in a continuous cycle.

The divergence of lithospheric plates generates new oceanic crust at spreading ridges, while convergence devours old oceanic crust in subduction zones. When two plates collide, the less buoyant oceanic crust subducts under continental crust. The lithosphere and the overriding oceanic crust recycle through the mantle to make new crust. The lithospheric plates act like rafts riding on a sea of molten rock, slowly carrying the continents around the surface of the globe.

LITHOSPHERIC PLATES

Earth's outer *shell* is fractured like a cracked egg into several large lithospheric plates (Fig. 44). The shifting plates range in size from a few hundred to several million square miles. They comprise the crust and the upper brittle mantle called the lithosphere. The lithosphere consists of the rigid outer layer of the mantle and underlies the continental and oceanic crust. The thickness of the lithosphere is about 60 miles under the continents and averages about 25 miles under the ocean.

Most continental rock originated when volcanoes stretching across the ocean were drawn together by plate tectonics. With the inclusion of continental margins and small shallow regions in the ocean, the continental crust covers about 45 percent of Earth's surface. It varies from 6 to 45 miles thick and rises on average about 4,000 feet above sea level. The thinnest parts of the continental crust lie below sea level on continental margins, and the thickest portions underlie mountain ranges.

The oceanic crust, by comparison, is considerably thinner and in most places is only 3 to 5 miles thick. Oceanic crust is only a small fraction of the age of continental crust, because the mantle at subduction zones spread around the world has consumed the older ocean floor. Perhaps as many as 20 oceans have come and gone during the last 2 billion years by the action of plate tectonics.

Figure 44 *The lithospheric plates that comprise Earth's crust.*

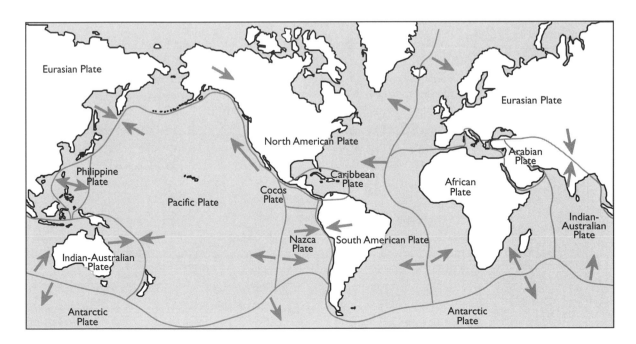

The lithosphere averages about 60 miles thick. It rides freely on the semimolten outer layer of the mantle, called the asthenosphere, between about 70 and 150 miles deep. This feature is necessary for the operation of plate tectonics. Otherwise, the crust would be jumbled-up slabs of rock.

Instead, eight major and about a half-dozen minor lithospheric plates carry the crust around on a sea of molten rock. The plates diverge at midocean ridges and converge at subduction zones, which are expressed on the ocean floor as deep-sea trenches. The trenches are regions where the plates are subducted into the mantle and remelted. The plates and oceanic crust are continuously recycled through the mantle. However, because of its greater buoyancy, the continental crust is rarely subducted.

An interesting feature about the crust geologists found quite by accident was that Scandinavia and parts of Canada are slowly rising nearly half an inch a year. Over the centuries, mooring rings on harbor walls in Baltic seaports have risen so far above sea level they could no longer be used to tie up ships. During the last ice age, the northern landmasses were covered with ice sheets up to 2 miles thick. Under the weight of the ice, North America and Scandinavia began to sink like an overloaded ship.

When the ice began to melt about 12,000 years ago, the extra weight was removed. As a result, the crust became lighter and began to rise (Fig. 45). In Scandinavia, marine fossil beds have risen more than 1,000 feet above sea level since the last ice age. The weight of the ice sheets depressed the landmass when the marine deposits were being laid down. When the ice sheets melted, the removal of the weight caused the landmass to rebound.

The lithospheric plates ride on a hot, pliable layer or asthenosphere in a manner similar to hard wax floating on melted wax. They carry the crust like drifting slabs of rock. The plates diverge at midocean spreading ridges and converge at subduction zones, lying at the edges of lithospheric plates. The lithospheric plates subduct into the mantle in a continuous cycle of crustal regeneration. Their constant interaction with each other shapes the surface of the planet. This structure of the upper mantle is important for the operation of plate tectonics, which is responsible for all geologic activity.

The plate boundaries are zones of active deformation that absorb the force of impact between nearly rigid plates. Throughout much of the world, clear geologic features, such as mountain ranges or deep ocean trenches, mark the boundaries between plates. These boundary zones vary from a few hundred feet where plates slide past each other at transform faults to several tens of miles at midocean ridges and subduction zones.

The divergent plate margins are midocean spreading ridges. These are regions where basalt welling up from within the upper mantle creates new oceanic crust as part of the process of seafloor spreading (Fig. 46). The midocean

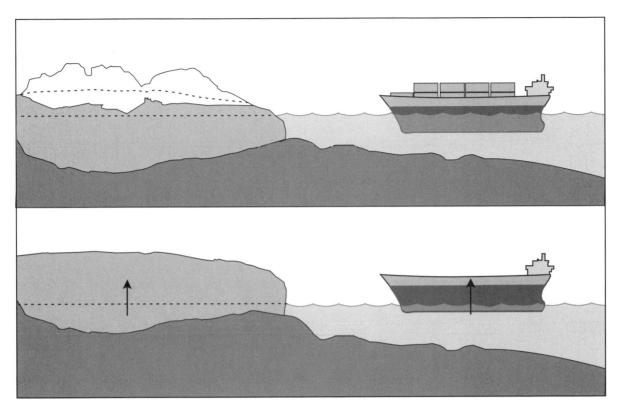

ridge system, which is not always found in the middle of the ocean, snakes 46,000 miles around the globe, making it the longest structure on Earth. The lateral plate margins are transform faults. These are regions where plates slide past each other accompanied by little or no tectonic activity, such as the upwelling of magma and the generation of earthquakes.

The convergent plate margins are the subduction zones represented by deep-sea trenches, where old oceanic crust sinks into the mantle to provide magma for volcanoes fringing the trenches. If tied end to end, the subduction zones would stretch completely around the world. The convergence rates between plates range from less than 1 inch to more than 5 inches per year, corresponding to the rates of plate divergence. However, subduction zones and associated spreading ridges on the margins of a plate do not operate at the same rates. This disparity causes the plates to travel across the surface of Earth. If subduction overcomes seafloor spreading, the lithospheric plate shrinks and eventually disappears altogether.

The oceanic plates thicken with age from a few miles thick, after formation at midocean spreading ridges, to more than 50 miles thick in the oldest ocean basins next to the continents. The depth at which an oceanic plate

Figure 45 *The principle of isostasy. Land covered with ice readjusts to the added weight like a loaded freighter. When the ice melts, the land is buoyed upward as the weight lessens.*

sinks as it moves away from a midocean spreading ridge varies by age. For example, a plate 2 million years old lies about 2 miles deep; a plate 20 million years old lies about 2.5 miles deep; and a plate 50 million years old lies about 3 miles deep.

A typical oceanic plate starts out thin. It thickens by the underplanting of new lithosphere from the upper mantle and by the accumulation of overlying sediment layers. The ocean floor at the summit of a midocean ridge consists almost entirely of hard basalt and acquires a thickening layer of sediments farther outward from the ridge crest. By the time the oceanic plate spreads out as wide as the Atlantic Ocean, the portion near continental margins where the sea is the deepest is about 60 miles thick. Eventually, the oceanic plate becomes so thick and heavy it can no longer remain on the surface. It then bends downward and subducts beneath a continent or another oceanic plate into Earth's interior (Fig. 47).

As the oceanic plate plunges into a subduction zone, it remelts and acquires new minerals from the mantle. This process provides the raw material for new oceanic crust as molten magma reemerges at volcanic spreading centers along midocean ridges. Sediments deposited onto the ocean floor and the water trapped between sediment grains are also caught in the subduction zones. However, the lower melting points and lesser density of these molten sediments cause them to rise toward the surface to supply nearby volcanoes with magma and recycled seawater.

Much more water is being subducted into Earth than emerges from subduction zone volcanoes. Heat and pressure act to dehydrate rocks of the

Figure 46 *Creation of oceanic crust at a spreading ridge.*

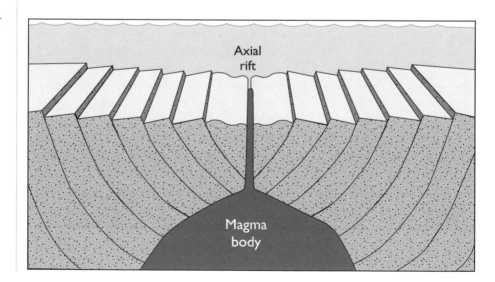

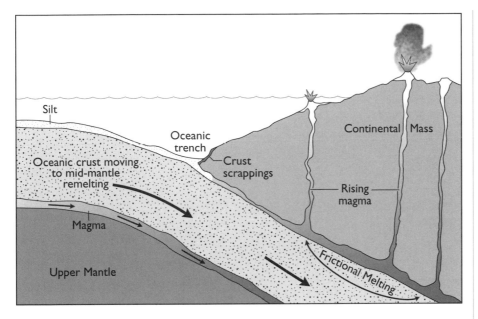

Figure 47 *The subduction of the ocean floor provides new molten magma for volcanoes that fringe the deep-sea trenches.*

descending plate. However, just where all the fluid goes has remained a mystery. Some fluid expelled from a subducting plate reacts with mantle rocks to produce low-density minerals that slowly rise to the seafloor. There they build mud volcanoes that erupt serpentine, an asbestos mineral formed by the reaction of water with olivine from the mantle, a silicate rich in iron and magnesium.

When the sediments and their contained seawater are caught between a subducting oceanic plate and an overriding continental plate, they are subjected to strong deformation, shearing, heating, and metamorphism. As the rigid lithospheric plate carrying the oceanic crust descends into Earth's interior, it slowly breaks up and melts. Over a period of millions of years, it is absorbed into the general circulation of the mantle. The subducted plate also supplies molten magma for volcanoes, most of which ring the Pacific Ocean and recycle chemical elements to Earth.

OCEANIC CRUST

The crust of the ocean is remarkable for its consistent thickness and temperature (Table 7). It averages about 4 miles thick and does not vary more than 20 degrees Celsius over most of the globe. Most oceanic crust is less than 4 percent of Earth's age and younger than 170 million years, with a mean age

---

TABLE 7   CLASSIFICATION OF THE EARTH'S CRUST

Environment	Crust Type	Tectonic Character	Thickness in Miles	Geologic Features
Continental crust overlying stable mantle	Shield	Very stable	22	Little or no sediment, exposed Precambrian rocks
	Midcontinent	Stable	24	
	Basin and range	Very unstable	20	Recent normal faulting, volcanism, and intrusion; high mean elevation
Continental crust overlying unstable mantle	Alpine	Very unstable	34	Rapid recent uplift, relatively recent intrusion; high mean elevation
	Island arc	Very unstable	20	High volcanism, intense folding and faulting
Oceanic crust overlying stable mantle	Ocean basin	Very stable	7	Very thin sediments overlying basalts, no thick Paleozoic sediments
Oceanic crust overlying unstable mantle	Ocean ridge	Unstable	6	Active basaltic volcanism, little or no sediment

---

of 100 million years. This is relatively young compared with continental crust, which is about 4 billion years old. Most of the seafloor has since disappeared into Earth's interior to provide the raw materials for the continued growth of the continents. The average density of continental crust is 2.7 times the density of water, compared with 3.0 for oceanic crust and 3.4 for the mantle. The difference in density buoys up the continental and oceanic crust.

Oceanic crust does not form as a single homogeneous mass. Instead, it comprises long, narrow ribbons laid side by side with fracture zones in between. The oceanic crust is comparable to a layer cake with four distinct strata (Fig. 48). The upper layer is pillow basalts, formed when lava extruded undersea at great depths. The second layer is of a sheeted-dike complex, consisting of a tangled mass of feeders that brought magma to the surface. The third layer is of gabbros, which are coarse-grained rocks that crystallized slowly under high pressure in a deep magma chamber. The fourth layer is of peridotites segregated from the mantle below. Gabbros containing higher amounts of silica solidify out of the basaltic melt and accumulate in the lower layer of the oceanic crust.

The same rock formation is found on the continents. This similarity led geologists to speculate that these formations were pieces of ancient oceanic crust called ophiolites, from the Greek word *ophis,* meaning "serpent." Ophiolites, so-named because of their mottled green color, date as far back as 3.6 billion years. These rocks were slices of ocean floor shoved up onto the continents by drifting plates. Therefore, ophiolites are among the best evidence for ancient plate motions.

Ophiolites are vertical cross sections of oceanic crust peeled off during plate collisions and plastered onto the continents. This action produced a linear formation of greenish volcanic rocks along with light-colored masses of granite and gneiss, common igneous and metamorphic rocks. Pillow lavas (Fig. 49), tubular bodies of basalt extruded undersea, are also found in the greenstone belts, signifying that these volcanic eruptions took place on the ocean floor. Many ophiolites contain ore-bearing rocks that are important mineral resources throughout the world.

Spreading ridges, where basalt oozes out of the mantle through rifts on the ocean floor, generate about 5 cubic miles of new oceanic crust every year. Some molten magma erupts as lava on the surface of the ridge through a system of vertical passages. Once at the surface, the liquid rock flows down the ridge and hardens into sheets or rounded forms of pillow lavas, depending on the rate of extrusion and the slope of the ridge. Periodically, lava overflows onto the ocean floor in gigantic eruptions, providing several square miles of new oceanic crust yearly. As the oceanic crust cools and hardens, it contracts and forms fractures through which water circulates.

Magma rising from the upper mantle extrudes onto the ocean floor and bonds to the edges of separating plates. Much of the magma solidifies

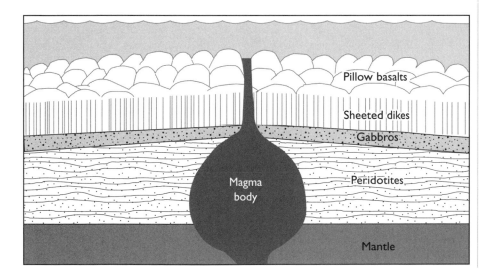

Figure 48 *The oceanic crust comprises a top layer of pillow basalts, a second layer of sheeted dikes, a third layer of gabbros, and an underlying layer of layered peridotites above the mantle.*

within the conduits above the magma chamber, forming massive vertical sheets called dikes that resemble a deck of cards standing on end. Individual dikes measure about 10 feet thick, stretch about 1 mile wide, and range about 3 miles long.

The asthenosphere is the fluid portion of the upper mantle, where rocks are semimolten or plastic, enabling them to flow slowly. After millions of years, the molten rocks reach the topmost layer of the mantle, or lithosphere. With a reduction of pressure within Earth, the rocks melt and rise through fractures in the lithosphere. As the molten magma passes through the lithosphere, it reaches the bottom of the oceanic crust, where it forms magma chambers that further press against the crust, which continues to widen the rift. Molten lava pouring out of the rift forms ridge crests on both sides and adds new material to the spreading ridge system (Fig. 50).

The mantle material below spreading ridges where new oceanic crust forms is mostly peridotite, a strong, dense rock composed of iron and magnesium silicates. As the peridotite melts on its journey to the base of the oceanic crust, a portion becomes highly fluid basalt, the most common magma erupted on the surface of Earth. About 5 cubic miles of basaltic magma are removed from the mantle and added to the crust every year. Most of this volcanism occurs on the ocean floor at spreading centers, where the oceanic crust pulls apart.

The oceanic crust, composed of basalts originating at spreading ridges and sediments washed off continents and islands, gradually increases density

Figure 49   *Pillow lava on the south bank of Webber Creek, Eagle District, Alaska.*

(Photo by E. Blackwelder, courtesy USGS)

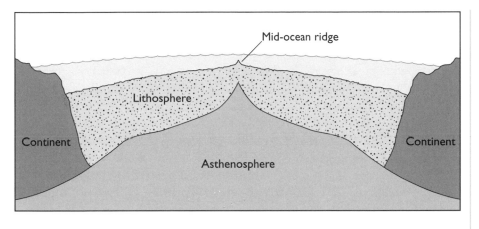

Figure 50 *Cross section of Earth beneath the Mid-Atlantic Ridge, which separated the New World from the Old World.*

and finally subducts into the mantle. On its way deep into Earth's interior, the lithosphere and its overlying sediments melt. The molten magma rises toward the surface in huge bubblelike structures called diapirs, from the Greek word *diapeirein,* meaning "to pierce." When the magma reaches the base of the crust, it provides new molten rock for magma chambers beneath volcanoes and granitic bodies called plutons such as batholiths (Fig. 51), which often form mountains. In this manner, plate tectonics is continuously changing and rearranging the face of Earth.

Figure 51 *A cutaway view of intrusive magma bodies that invade Earth's crust and erupt on the surface as volcanoes.*

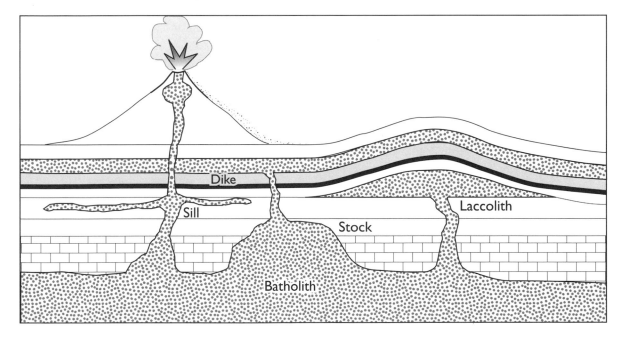

THE ROCK CYCLE

The development of the theory of plate tectonics has led to a greater under-standing of the geochemical carbon cycle or simply rock cycle. The rock cycle is extremely crucial for keeping the planet alive both geologically and biolog-ically. The recycling of carbon through the geosphere makes Earth unique among planets. This is evidenced by the fact that the atmosphere contains large amounts of oxygen. Without the carbon cycle, oxygen would have long since been buried in the geologic column, comprising the sedimentary rocks that make up the crust. Fortunately, plants replenish oxygen by utilizing carbon dioxide, which plays a critical role as a primary source of carbon for photo-synthesis and therefore provides the basis for all life.

The entire volume of the world's oceans circulates through the crust at spreading ridges every 10 million years. This circulation is approximately equivalent to the annual flow of the Amazon, the world's largest river. This action accounts for the unique chemistry of seawater and for the efficient thermal and chemical exchanges between the crust and the ocean. The mag-nitude of some of these chemical exchanges is comparable in volume to the input of elements into the oceans by the world's rivers, which carry materials weathered from the continents. The most important of these chemical ele-ments is carbon, which controls many life processes on the planet.

When the seafloor subducts into Earth's interior, the intense heat of the mantle drives out carbon dioxide from carbonaceous sediments. The molten rock, with its contingent of carbon dioxide, flows upward through the mantle and fills the magma chambers underlying volcanoes and spreading ridges. The consequent eruption of volcanoes and the flow of molten rock from midocean ridges resupplies the atmosphere with new carbon dioxide, making Earth one great carbon dioxide recycling plant (Fig. 52).

The geochemical carbon cycle is the transfer of carbon within Earth and involves the interactions between the crust, ocean, atmosphere, and life (Table 8). Many aspects of this important cycle were understood around the turn of the 20th century, notably by the American geologist Thomas Chamberlain and chemist Harold Urey, who developed the theory. However, only in the last few years has the geochemical carbon cycle been placed within the more comprehensive framework of plate tectonics.

The biologic carbon cycle is only a small component of this cycle. It is the transfer of carbon from the atmosphere to vegetation by photosynthesis and then the returning of carbon to the atmosphere when plants respire or decay. Only about one-third of the chemical elements, mostly hydrogen, oxy-gen, carbon, and nitrogen, comprising most of the elements of life, are recy-cled biologically. The vast majority of carbon is not stored in living tissue but locked up in sedimentary rocks. Even the amount of carbon contained in fos-

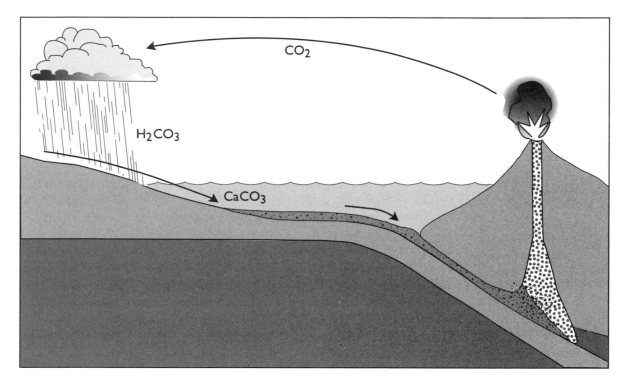

sil fuels is meager by comparison. The combustion of large quantities of fossil fuels and the destruction of the world's forests is transferring more carbon to the atmosphere than can be disposed of in the ocean, which holds by far the largest store of carbon dioxide. The burning of carbon-based fuels could dramatically influence the climate through the greenhouse effect.

The biosphere, the living portion of the planet, plays a very important role in the cycling of carbon. The creation and decomposition of peat bogs might have been responsible for most changes in levels of atmospheric carbon dioxide during the past two glaciations. The bogs have accumulated upward of 250 billion tons of carbon in the last 10,000 years since the end of the last ice age, mostly in the temperate zone of the Northern Hemisphere. Over geologic time, progressively more land has drifted into latitudes where large quantities of carbon are stored as peat. During the last million years, glaciations have gradually remolded large parts of the Northern Hemisphere into landforms more suitable for peat bog formation in wetlands.

Carbon dioxide presently comprises about 365 parts per million of the air molecules in the atmosphere, amounting to about 800 billion tons of carbon. It is one of the most important greenhouse gases, which traps solar heat that would otherwise escape into space. Carbon dioxide, therefore, operates

Figure 52 *The geochemical carbon cycle. Carbon dioxide in the form of bicarbonate is washed off the land and enters the ocean, where organisms convert it to carbonate sediments, which are thrust into the mantle, become part of the magma, and escape into the atmosphere from volcanoes.*

TABLE 8   THE AMOUNT OF CARBON RELATIVE TO LIFE

Source	Relative Amount
Calcium carbonate in sedimentary rocks	60,000
Ca-Mg carbonate in sedimentary rocks	45,000
Sedimentary organic matter in the remains of animal tissues	25,000
Bicarbonate and carbonate dissolved in ocean	75
Coal and petroleum	7
Soil humus	5
Atmospheric carbon dioxide	1.5
All living plants and animals	1

---

somewhat like a thermostat to regulate the temperature of the planet. Since it plays such a critical role in regulating Earth's temperature, major changes in the carbon cycle could have profound climatic effects. If the carbon cycle removes too much carbon dioxide, Earth cools. If the carbon cycle generates too much carbon dioxide, Earth warms. Therefore, even slight changes in the carbon cycle could have a considerable influence on the climate.

The ocean plays a crucial role in the carbon cycle by regulating the level of carbon dioxide in the atmosphere. In the upper layers of the ocean, the concentration of gases is in equilibrium with the atmosphere. The mixed layer of the ocean (Fig. 53), within the upper 300 feet, contains as much carbon dioxide as the entire atmosphere. The gas dissolves into seawater mainly by the agitation of surface waves. Without marine photosynthetic organisms to absorb dissolved carbon dioxide, much of the ocean's reservoir of this gas would escape into the atmosphere, more than tripling the present content and causing a runaway greenhouse effect.

A major portion of the carbon in the ocean comes from the land. Atmospheric carbon dioxide combines with rainwater to form carbonic acid. The acid reacts with surface rocks, producing dissolved calcium and bicarbonate, which are carried by streams to the sea. Marine organisms use these substances to build calcium carbonate skeletons and other supporting structures. When the organisms die, their skeletons sink to the bottom of the ocean, where they dissolve in the deep abyssal waters. Because of its large volume, the abyss holds the largest reservoir of carbon dioxide, containing 60 times more carbon than the atmosphere.

Sediments on the ocean floor and on the continents store most of the carbon in carbonate rocks. In shallow water, the carbonate skeletons build deposits of limestone (Fig. 54), which buries carbon dioxide in the geologic column. The burial of carbonate is responsible for about 80 percent of the carbon deposited onto the ocean floor. The carbon locked up in carbonate minerals in the upper crust is estimated at 800 trillion tons. The remainder of the carbonate originates from the burial of dead organic matter washed off the continents.

In this respect, marine life acts as a pump to remove carbon dioxide from the atmosphere and the ocean's surface waters and store it in the deep sea. The faster this biologic pump works, the more carbon dioxide that is removed from the atmosphere, with the rate determined by the amount of nutrients in the ocean. A reduction of nutrients slows the pump, returning deep-sea carbon dioxide to the atmosphere.

Half the carbonate transforms back into carbon dioxide, which returns to the atmosphere, mostly by upwelling currents in the tropics. Therefore, the concentration of atmospheric carbon dioxide is highest near the equator. If not for this process, in a mere 10,000 years all carbon dioxide would be removed from the atmosphere. Without the return of carbon dioxide to the atmosphere, photosynthesis would cease and all life would end.

The deep water, which represents about 90 percent of the ocean's volume, circulates very slowly with a residence time (time in place) of about 1,000 years. It comes into direct contact with the atmosphere only in the polar regions. Thus, the deep water's absorption of carbon dioxide is limited. The

Figure 53 *Turbulence in the upper layers of the ocean induces the mixing of temperatures, nutrients, and gases.*

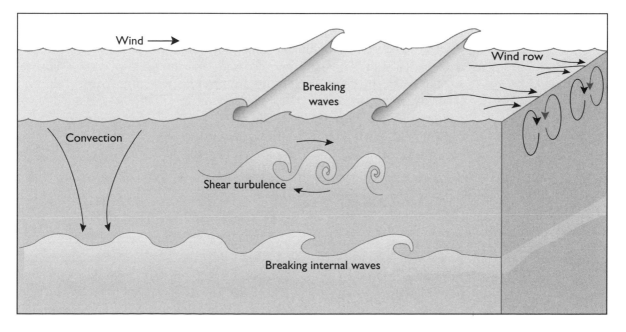

Figure 54 *The formation of limestone from carbonaceous sediments deposited onto the ocean floor.*

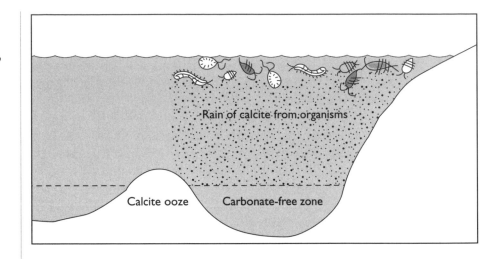

abyss receives most of its carbon as shells of dead organisms and fecal matter that sink to the ocean bottom.

Volcanic activity on the ocean floor and on the continents plays a vital role in restoring the carbon dioxide content of the atmosphere. Carbon dioxide escapes from carbonaceous sediments that melt in Earth's interior to provide new magma. The molten magma along with volatiles including water and carbon dioxide rise to the surface to feed magma chambers beneath midocean ridges and volcanoes. When the volcanoes erupt, carbon dioxide releases from the magma and returns to the atmosphere, completing the cycle.

The cycling of carbon through land, ocean, and atmosphere is significant in determining potential climate change. Atmospheric scientists are particularly interested in knowing what role rising concentrations of carbon dioxide, an important greenhouse gas, might play in warming Earth's climate. Researchers do not know how Earth stores much of the carbon dioxide released by human activity or whether the planet's storage capacity for the gas can continue indefinitely. Therefore, decisions about any future actions regarding management of the carbon cycle will require additional scientific inquiry.

OCEAN BASINS

Ocean basins are the largest and deepest depressions on Earth. The ocean floor lies much deeper below sea level than the continents rise above it. If the oceans were completely drained of water, the planet would look much like the rugged surface of Venus, which lost its oceans eons ago. The deepest parts of the dry seabed would lie several miles below the surrounding continental margins. The floor of the desiccated ocean would be traversed by the longest mountain ranges and fringed in places by the deepest trenches. Vast empty basins would divide the continents, which would stand out like thick slabs of rock.

Most of the seawater surrounding the continents lies in a single great basin in the Southern Hemisphere, which is nine-tenths ocean. It branches northward into the Atlantic, Pacific, and Indian basins in the Northern Hemisphere, where most of the continental landmass exists. The Arctic Ocean is a nearly landlocked sea connected to the Atlantic and Pacific by narrow straits. The Bering Sea separates Alaska and Asia by only 56 miles at its narrowest point (Fig. 55). About 20 million years ago, a ridge near Iceland subsided, allowing cold water from the recently formed Arctic Ocean to surge into the Atlantic, giving rise to the oceanic circulation system in existence today (Table 9).

The oceans expand across some 70 percent of Earth's surface. They cover an area of about 140 million square miles with more than 300 million cubic

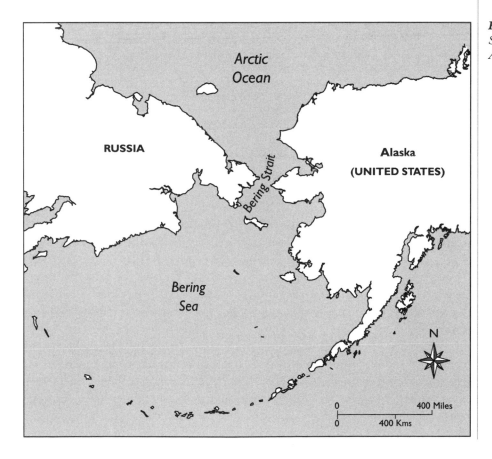

Figure 55 *The Bering Strait between Alaska and Asia.*

TABLE 9   HISTORY OF THE DEEP CIRCULATION OF THE OCEAN

Age (millions of years ago)	Event
3	An ice age overwhelms the Northern Hemisphere.
3–5	Arctic glaciation begins.
15	The Drake Passage is open; the circum-Antarctic current is formed. Major sea ice forms around Antarctica, which is glaciated, making it the first major glaciation of the modern Ice Age. The Antarctic bottom water forms. The snow limit rises.
25	The Drake Passage between South America and Antarctica begins to open.
25–35	A stable situation exists with possible partial circulation around Antarctica. The equatorial circulation is interrupted between the Mediterranean Sea and the Far East.
35–40	The equatorial seaway begins to close. There is a sharp cooling of the surface and of the deep water in the south. The Antarctic glaciers reach the sea with glacial debris in the sea. The seaway between Australia and Antarctica opens. Cooler bottom water flows north and flushes the ocean. The snow limit drops sharply.
> 50	The ocean could flow freely around the world at the equator. Rather uniform climate and warm ocean occur even near the poles. Deep water in the ocean is much warmer than it is today. Only alpine glaciers exist on Antarctica.

miles of seawater. About 60 percent of the planet is covered by water no less than 1 mile deep. The average depth is about 2.3 miles. The midocean ridges lie at an average depth of 1.5 miles, and the ocean bottom slopes away on both sides to a depth of about 3.5 miles. In the Pacific basin, the ocean is up to 7 miles deep—the lowest point on Earth.

With only marine-borne sedimentation and no bottom currents to stir up the seabed, an even blanket of material would settle onto the original volcanic floor of the oceans. Instead, the rivers of the world contribute a substantial amount of the sediments deposited onto the deep ocean floor. The largest rivers of North and South America empty into the Atlantic, which receives considerably more river-borne sediments than does the Pacific. The burial of organic material also greatly aids the formation of offshore petroleum reserves.

Because the Atlantic is smaller and shallower than the Pacific, its marine sediments are buried more rapidly and are therefore more likely to survive than those in the Pacific. The floor of the Atlantic accumulates considerably more sediments than that of the Pacific, amounting to a rate of about an inch every 2,500 years. The deep-ocean trenches around the Pacific trap much of the material reaching its western edge, where it subducts into the mantle.

Strong near-bottom currents redistribute sediments in the Atlantic on a greater scale than in the Pacific. Abyssal storms with powerful currents occasionally sweep patches of ocean floor clean of sediments and deposit the debris elsewhere. On the western side of the ocean basins, periodic undersea storms skirt the foot of the continental rise and transport huge loads of sediment, dramatically modifying the seafloor. The scouring of the seabed and deposition of thick layers of fine sediments result in much more complex marine geology than that developed simply from a constant rain of sediments from above.

SUBMARINE CANYONS

The ocean floor presents a rugged landscape unmatched anywhere else on Earth. Chasms dwarfing even the largest continental canyons plunge to great depths. The Romanche Fracture Zone (Fig. 56) is a deep trench offsetting the axis of the Mid-Atlantic Ridge eastward nearly 600 miles. The bottom of the

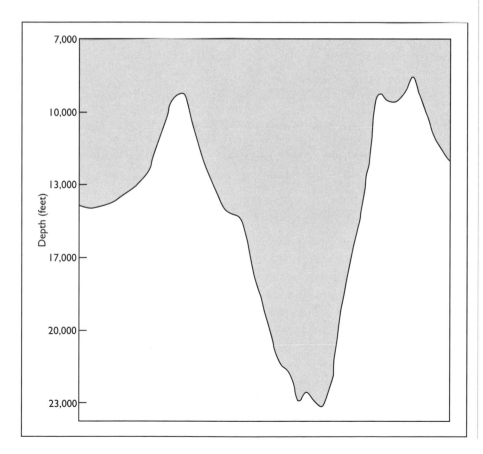

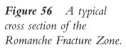

Figure 56 *A typical cross section of the Romanche Fracture Zone.*

Figure 57 *The extended shoreline during the height of the Ice Age.*

Romanche trench is in places 5 miles below sea level. The highest parts of the ridges on either side of the trench are less than 1 mile below sea level, giving a vertical relief of four times that of the Grand Canyon.

Rivers emptying into the sea carved out many submarine canyons. They eroded the exposed seabed when the sea level lowered dramatically during the last ice age. At the height of glaciation, about 10 million cubic miles of Earth's water were held in the continental ice sheets, which covered about one-third of the land surface with an ice volume three times greater than at present. The accumulated ice dropped the level of the ocean by about 400 feet, advancing the shoreline hundreds of miles seaward (Fig. 57). The coastline of the eastern seaboard of the United States extended about halfway to the edge of the continental shelf, which runs eastward more than 600 miles. The drop in sea level exposed land bridges and linked continents.

Numerous canyons slice through the continental shelf beneath the Bering Sea between Alaska and Siberia. About 75 million years ago, continental movements created the broad Bering shelf rising 8,500 feet above the deep ocean floor. The shelf was exposed as dry land several times during the ice ages when sea levels dropped hundreds of feet and terrestrial canyons cut deep into the shelf. When the ocean refilled again at the end of the last ice age, massive landslides and mudflows swept down steep slopes on the shelf's edge, gouging out 1,400 cubic miles of sediments and rocks.

A step resembling a sea cliff on the continental shelf off the eastern United States has been traced for nearly 200 miles. It appears to represent the former Ice Age coastline, now completely submerged under seawater. The massive continental glaciers that sprawled over much of the Northern Hemisphere held enough water to lower the sea by several hundred feet. When the glaciers melted, the sea returned to near its present level. Submarine canyons carved into bedrock 200 feet below sea level can be traced to rivers on land.

Several submarine canyons slice through the continental margin and ocean floor off eastern North America (Fig. 58). Submarine canyons on continental shelves and slopes have many identical features as river canyons, and some rival even the largest on the continents. High, steep walls and an irregular floor that slopes continually outward characterize them. The canyons are up to 30 miles and more in length, with an average wall height of about 3,000 feet. Some submarine canyons were carved out of the ocean floor by ordinary river erosion during a time when sea levels were much lower than they are today. The Great Bahamas Canyon is one of the largest submarine canyons, with a wall height of 14,000 feet, more than twice as deep as the Grand Canyon.

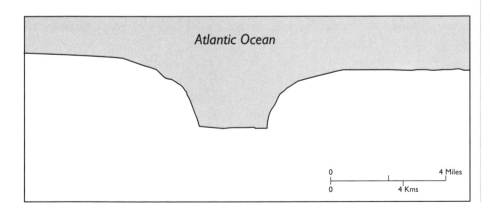

Figure 58 *The Midocean Canyon in the Newfoundland basin.*

(Photo by R. M. Pratt, courtesy USGS)

Rivers flowing across the exposed land gouged out several submarine canyons in the ocean floor when sea levels were much lower than today. Many submarine canyons have heads near the mouths of large rivers. Some submarine canyons extend to depths of more than 2 miles, much too deep for a terrestrial river origin. They formed instead by undersea slides, which carve out deep gashes in the ocean floor.

The Mediterranean Sea appears to have almost completely dried up 6 million years ago due to the uplift of the Strait of Gibraltar closing off the Atlantic. Its seafloor became a desert basin more than 1 mile below the surrounding continental plateaus. Rivers draining into the desiccated basin gouged out deep canyons. A deep, sediment-filled gorge follows the coarse of the Rhône River in southern France for more than 100 miles and extends to a depth of 3,000 feet below the surface, where the river drains into the Mediterranean Sea. Under the sediments of the Nile Delta is buried a mile-deep canyon that can be traced as far south as Aswan 750 miles away. It is comparable in size to the largest canyons on the continents.

Submarine slides move rapidly down steep continental slopes and are responsible for excavating deep submarine canyons. The surface of the slopes is covered mainly with fine sediments swept off the continental shelves by submarine slides. The slides consist of sediment-laden water that is denser than the surrounding seawater. The turbid water moves swiftly along the ocean floor, eroding the soft bottom material. These muddy waters, called turbidity currents, move down steep slopes and play a major role in shifting the sands of the deep sea (Fig. 59).

The continental slope marks the boundary between the shallow continental shelf and the deep ocean. It inclines as much as 60 to 70 degrees and plunges downward for thousands of feet. Sediments that reach the edge of the continental shelf slide off the continental slope by the pull of gravity. Huge masses of sediment cascade down the continental slope by gravity slides that can gouge out steep submarine canyons and deposit great heaps of sediment. They are often as catastrophic as terrestrial slides and can move massive quantities of sediment downslope in a matter of hours.

Some portions of the continental slope around the United States, including regions off the mid-Atlantic states, western Florida, Louisiana, California, and Oregon, have extremely steep slopes. The steepest of the slopes, off western Florida, has a near-vertical escarpment. The seafloor has such an extreme incline here because groundwater peculating down to the base of the continental slope erodes the rock, causing the bottom to collapse. In contrast, narrow canyons cut into the side of the slope off New Jersey, giving it the appearance of an eroded mountain range. The Louisiana seafloor bears the marks of numerous craters created by the eruption of buried salt deposits, giving the seabed the look of a moonscape.

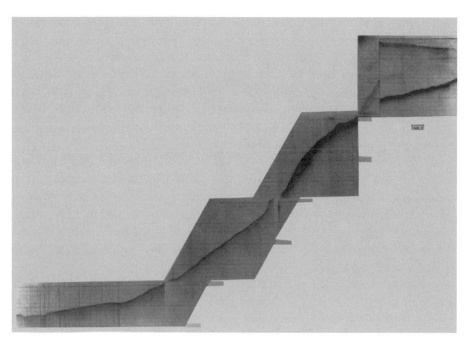

Figure 59 Sonar profile down the continental slope south of Nantucket, showing slumped debris from 5,000 to 7,500 feet deep.

(Photo by R. M. Pratt, courtesy USGS)

MICROPLATES AND TERRANES

During the initial breakup of Pangaea in the early Jurassic period, the Pacific plate was hardly bigger than the United States. Now it is the largest plate in the world. About 190 million years ago, the Pacific plate might have begun as a tiny microplate, a small block of oceanic crust that sometimes lies at the junction between two or three major plates. The rest of the ocean floor consisted of other unknown plates that have long since disappeared as the Pacific plate continued to grow. This is why no oceanic crust is older than Jurassic in age.

A microplate about the size of Ohio sits at the junction of the Pacific, Nazca, and Antarctic plates in the South Pacific about 2,000 miles west of South America. Seafloor spreading along the boundary zone between the plates adds new oceanic crust onto their edges, causing the plates to diverge. The different rates of seafloor spreading have caused the microplate at the hub of the spreading ridges, which fan out like the spokes of a bicycle wheel, to rotate one-quarter turn clockwise in the last 4 million years. A similar microplate near Easter Island to the north has spun nearly 90 degrees over the last 3 to 4 million years, suggesting that most microplates behave in this manner.

Three lithospheric plates bordering the Pacific Ocean—the Nazca, Antarctic, and South American plates—come together in an unusual triple junction. The first two plates spread apart along a boundary called the Chile Ridge off the west coast of South America, similar to the way the Americas drift away from Eurasia and Africa along the Mid-Atlantic Ridge. The Chile Ridge lies off the Chilean continental shelf at a depth of more than 10,000 feet. Along its axis, magma rises from deep within Earth and piles up into mounds, forming undersea volcanoes.

The Nazca plate moving northeast subducts beneath the westward-moving South American plate at the Peru-Chile Trench. The eastern edge of the Nazca plate is subducting at a rate of about 50 miles every million years, which is faster than its western edge is growing. An analogy would be an escalator whose top is moving toward the bottom, consuming the rising steps in between. In essence, the Chile Trench is eating away at the Chile Ridge, which will eventually disappear.

Several times in the past 170 million years, other plates and their associated spreading centers have vanished beneath the continents that surround the Pacific basin. This had a substantial impact on the coastal geology. A high degree of geologic activity around the rim of the Pacific basin produced virtually all the mountain ranges facing the Pacific and the island arcs along its perimeter. Most of the terranes (wandering blocks of crust) that collided with the North American continent originated in the Pacific.

Terranes are patches of oceanic crust originating from faraway sources that have been shoved up onto the continents and assembled into geologic collages. The term terrane should not be confused with terrain, which means landform. Terranes exist in a variety of shapes and sizes. They range from small slivers to subcontinents such as India, which is a single great terrane. Terranes are distinct from their geologic surroundings and are usually bounded by faults. The boundary between two or more terranes is called a suture zone. The composition of terranes generally resembles that of an oceanic island or plateau. However, some comprise a consolidated conglomerate of pebbles, sand, and silt that accumulated in an ocean basin between colliding crustal fragments.

Most terranes created on an oceanic plate are elongated bodies that deformed when colliding with a continent, which subjects them to crustal movements that modify their shape. The assemblage of terranes in China is being stretched and displaced in an east-west direction due to the continuing pressure that India exerts on southern Asia as it raises the Himalayas (Fig. 60). Granulite terranes are high-temperature metamorphic belts, which formed in the deeper parts of continental rifts. They also comprise the roots of mountain belts formed by continental collision, such as the Alps and Himalayas.

North of the Himalayas lies a belt of ophiolites, which appears to mark the boundary between the sutured continents. Ophiolites are slices of ocean floor shoved up onto the continents by drifting plates and date as old as 3.6 billion years. Terrane boundaries are commonly marked by ophiolite belts, consisting of marine sedimentary rocks, pillow basalts, sheeted dike complexes, gabbros, and peridotites.

Suspect terranes, named so because of their exotic origins, are fault-bounded blocks with geologic histories distinct from those of neighboring terranes and of adjoining continental masses. They range in age from less than 200 million years old to more than 1 billion years old. Suspect terranes were displaced over great distance before finally being accreted to a continental margin. Some North American suspect terranes have a western Pacific origin and were displaced thousands of miles to the east.

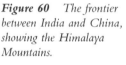

Figure 60 *The frontier between India and China, showing the Himalaya Mountains.*

(Photo courtesy NASA)

Figure 61 *Radiolarians were marine planktonic protozoans.*

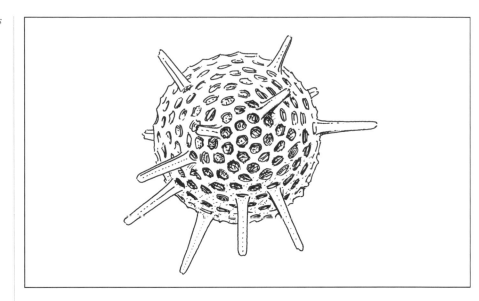

The actual distances terranes can travel varies considerably. Basaltic seamounts that accreted to the margin of Oregon moved from nearby off-shore. Similar rock formations around San Francisco, California, however, came from halfway across the Pacific Ocean. The city itself is built on three different and distinct rock units. At their usual rate of travel, terranes could make a complete circuit of the globe in only about half a billion years. Different species of fossil radiolarians (Fig. 61), which are marine protozoans with skeletons of silica and were abundant from about 500 million to 160 million years ago, determine the age of the terranes. They also define specific regions of the ocean where the terranes originated.

Much of western North America was assembled from oceanic island arcs and other crustal debris skimmed off the Pacific plate as the North American plate headed westward. Until about 250 million years ago, the western edge of North America ended near present-day Salt Lake City. Over the last 200 million years, North America has expanded by more than 25 percent during a major pulse of crustal growth. Northern California is a jumble of crust assembled some 200 million years ago. A nearly complete slice of ocean crust at least 2.7 billion years old shoved up onto the continents by drifting plates sits in the middle of Wyoming.

Many terranes that comprise western North America have rotated clockwise as much as 70 degrees or more, with the oldest terranes having the greatest degree of rotation. Terranes created on an oceanic plate retain their shapes until they collide and accrete to a continent. They are then subjected to crustal movements that modify their overall dimensions.

Alaska is an agglomeration of terranes comprising pieces of an ancient ocean that preceded the Pacific, called the Panthalassa. The terranes that make up the Brooks Range (Fig. 62), the spine of northern Alaska, are great sheets stacked one on top of another. The entire state is an assemblage of some 50 terranes set adrift over the past 160 million years by the wanderings and collisions of crustal plates, parts of which are still arriving from the south. Western California has been drifting northward for millions of years. In another 50 million years, it will wander as far north as Alaska, adding another piece to the puzzle.

The Alexander terrane, comprising a large portion of the Alaskan panhandle, began as part of eastern Australia some 500 million years ago. About 375 million years ago, it broke free from Australia, traversed the Pacific Ocean, stopped briefly at the coast of Peru, sliced past California and swiped part of the Mother Lode gold belt, and crashed into the North American continent around 100 million years ago.

The accreted terranes also played a major role in the creation of mountain chains along convergent continental margins. For example, the Andes appeared to have been raised by the accretion of oceanic plateaus along the continental margin of South America. Along the mountain ranges in western

Figure 62 *Steeply dipping Paleozoic rocks of the Brooks Range, near the head of the Itkillik River east of Anaktuvuk Pass, Northern Alaska.*

(Photo by J. C. Reed, courtesy USGS)

North America, the terranes are elongated bodies due to the slicing of the crust by a network of northwest-trending faults. One of these is the San Andreas Fault in California (Fig. 63), which has undergone some 200 miles of displacement in the last 25 million years.

After discussing the activity of the oceanic crust, the next chapter examines the effects of seafloor spreading and plate subduction.

4

RIDGES AND TRENCHES
UNDERSEA MOUNTAINS AND CHASMS

This chapter examines the major geologic structure on the seabed. The ocean floor offers a rugged landscape unmatched by anywhere on Earth's surface. Vast undersea mountain ranges, much more extensive than those on the continents, crisscross the ocean stretches. A continuous system of midocean ridges girdles the planet. It is by far the longest geologic structure on the planet. Although deeply submerged, the midocean ridge system is easily the most dominant feature, extending over a larger area than all major terrestrial mountain ranges combined.

The subduction of the lithosphere in deep-sea trenches plays a fundamental role in global tectonics and accounts for powerful geologic forces that continuously shape the surface of Earth. Major mountain ranges and most volcanoes are associated with the subduction of lithospheric plates. The subduction of the oceanic crust into the mantle produces strain in the descending lithosphere, causing powerful earthquakes to rumble across the landscape and seascape.

THE MIDOCEAN RIDGES

The shifting lithospheric plates create new oceanic crust in a continuous cycle of crustal rejuvenation. The subducting lithosphere circulates through the mantle and reemerges as magma at a dozen or so midocean ridges around the world, generating more than half of Earth's crust. The addition of new basalt on the ocean floor is responsible for the growth of the lithospheric plates upon which the continents ride.

A large part of this activity takes place in the middle of the Atlantic Ocean, where molten rock welling up from the upper mantle generates new sections of oceanic crust. The floor of the Atlantic acts like two opposing conveyor belts. The rollers are convection loops in the upper mantle, transporting oceanic crust in opposite directions outward from its point of origin at the Mid-Atlantic Ridge.

The spreading ridge system runs from Bovet Island, about 1,000 miles north of Antarctica, to Iceland, itself a surface expression of the Mid-Atlantic Ridge (Fig. 64). Extensive volcanism gives rise to volcanic islands such as Surtsey (Fig. 65), located 7 miles to the south. The midocean ridge is a string of volcanic seamounts, created by molten magma upwelling from within the mantle. Running down the middle of the ridge crest is a deep trough, as though a giant crack were carved in the ocean's crust. This trough reaches 4 miles in depth and up to 15 miles wide, making it the greatest chasm on Earth.

The submerged mountains and undersea ridges form a continuous chain 46,000 miles long (Fig. 66). It is by far the largest structure on the planet, surpassing in scale all mountain ranges on land. The mountainous belt is several hundred miles wide and up to 10,000 feet above the ocean floor. When starting from the Arctic Ocean, the ridge spans southward across the Atlantic basin; continues around Africa, Asia, and Australia; runs under the Pacific Ocean; and terminates at the west coast of North America.

Figure 64 *Iceland straddles the Mid-Atlantic Ridge.*

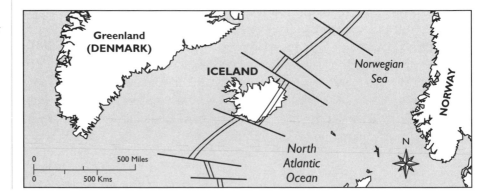

Figure 65 *Birth of the new Icelandic island, Surtsey, in November 1963, located 7 miles south of Iceland.*

(Photo courtesy U.S. Navy)

The ocean floor at the crest of the ridge consists mainly of basalt, the most common magma erupted on the surface of Earth. About 5 cubic miles of new basalt are added to the crust annually, mostly on the ocean floor at spreading ridges. With increasing distance from the crest, a thickening layer of sediments shrouds the bare volcanic rock. As the two newly separated plates move away from the rift, material from the asthenosphere adheres to their edges to form new lithosphere. The lithospheric plate thickens as it propagates from a midocean rift system, causing the plate to sink deeper into the mantle. This is why the seafloor near the continental margins surrounding the Atlantic basin is the deepest part of the Atlantic Ocean.

Intense seismic and volcanic activity along the midocean ridges manifest as a high heat flow from Earth's interior. Molten magma originating from the mantle rises through the lithosphere and adds new basalt to both sides of the ridge crest. The greater the flow of magma, the more rapid is the seafloor spreading and the lower the relief. The spreading ridges in the Pacific Ocean are more active than those in the Atlantic and therefore are less elevated. Rapidly spreading ridges do not achieve the heights of slower ones simply because the magma does not have sufficient time to pile up into tall heaps.

The axis of a slow-spreading ridge is characterized by a rift valley several miles deep and about 10 to 20 miles wide.

A set of closely spaced fracture zones dissects the Mid–Atlantic Ridge in the equatorial Atlantic. The largest of these structures is the Romanche Fracture Zone (Fig. 67). It offsets the axis of the ridge in an east–west direction by nearly 600 miles. The floor of the Romanche trench is as much as 5 miles below sea level. The highest parts of the ridges on either side of the trench are less than 1 mile below sea level. This provides a vertical relief of four times that of the Grand Canyon.

The shallowest portion of the ridge is capped with a fossil coral reef, suggesting it was above sea level some 5 million years ago. Many similar and equally impressive fracture zones span the area, culminating in a sequence of troughs and transverse ridges several hundred miles wide. The resulting terrain is unmatched in size and ruggedness anywhere else in the world.

In the Pacific Ocean, a rift system called the East Pacific Rise stretches 6,000 miles from the Antarctic Circle to the Gulf of California. It lies on the eastern edge of the Pacific plate, marking the boundary between the Pacific and Cocos plates. It is the counterpart of the Mid–Atlantic Ridge and a member of the world's largest undersea mountain chain. The rift system is a network of midocean ridges, which lie mostly at a depth of about 1.5 miles. Each rift is a narrow fracture zone, where plates of the oceanic crust diverge at an average rate of about 5 inches a year. This results in less topographical relief on the ocean floor. The active tectonic zone of a fast spreading ridge is usually quite narrow, generally less than 4 miles wide.

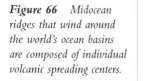

Figure 66 *Midocean ridges that wind around the world's ocean basins are composed of individual volcanic spreading centers.*

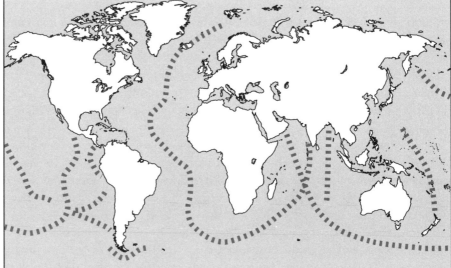

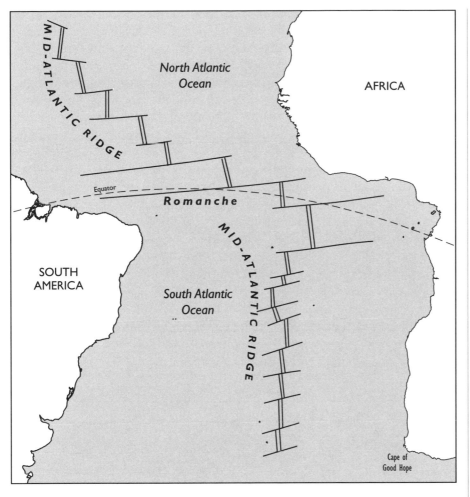

THE HEAT ENGINE

All geologic activity that continuously shapes the surface of Earth is an outward expression of the great heat engine in the interior of the planet. The motion of the mantle churning over ever so slowly below the crust brings heat from the core to the surface by convection loops (Fig. 68), the main driving force behind plate tectonics. Convection is the motion within a fluid medium that results from a difference in temperature from the bottom to the top. The core transfers heat to mantle rocks, whose increased buoyancy causes them to rise to the surface.

Convection currents and mantle plumes transport molten magma to the underside of the lithosphere, which is responsible for most of the volcanic activity on the ocean floor and on the continents. Most mantle plumes originate from within the mantle. However, some arise from the very bottom of the mantle, making it a huge, bubbling pot stirred throughout its entire depth.

The formation of molten rock and the rise of magma to the surface results from an exchange of heat within the planet's interior. Fluid rocks in the mantle acquire heat from the core, ascend, dissipate heat to the lithosphere, cool, and descend back to the core to be reheated. The mantle currents travel very slowly, completing a single convection loop in several hundred million years.

Earth is steadily losing heat from its interior to the surface through the lithosphere. About 70 percent of this heat loss results from seafloor spreading. Most of the rest is due to volcanism at subduction zones (Fig. 69). Lithospheric plates created at spreading ridges and destroyed at subduction zones are the final products of convection currents in the mantle.

Most of the mantle's heat originates from internal radiogenic sources. The rest comes from the core, which has retained much of its original heat since the early accretion of Earth some 4.6 billion years ago. The temperature difference between the mantle and the core is nearly 1,000 degrees Celsius. Material from the mantle might be mixing with the fluid outer core to form a distinct layer on the surface that could block heat flowing from the core to the mantle and interfere with mantle convection.

Figure 68 Convection currents in the mantle move the continents around Earth.

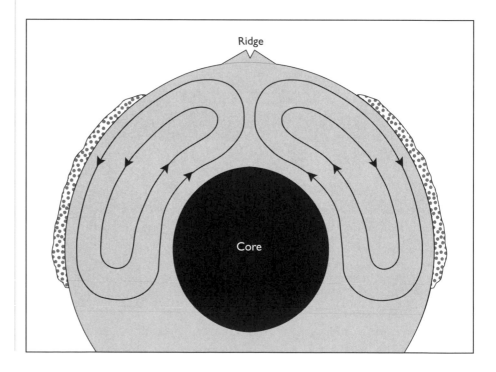

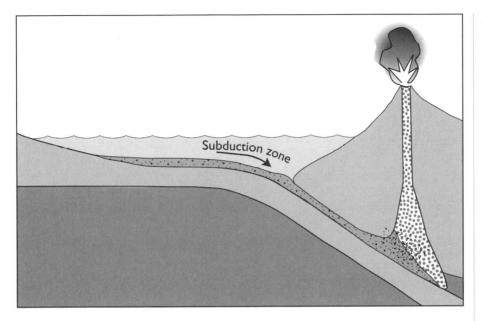

Figure 69 *The subduction of a lithospheric plate into the mantle supplies volcanoes with molten magma.*

Heat transferred from the mantle to the asthenosphere causes convection currents to rise and travel laterally when reaching the underside of the lithosphere. Upon giving up their heat energy to the lithosphere, the currents cool and descend back into the mantle in a manner similar to air currents in the atmosphere. If any cracks or areas of weakness occur in the lithosphere, the convective currents spread the fissures wider apart to form undersea spreading ridges in the ocean and rift systems on the continents (Fig. 70). Here the largest proportion of Earth's interior heat is lost to the surface as magma flows out of the rift zones.

The mantle rocks are churning over very slowly in large-scale convection loops. They travel only a couple inches a year, about the same as plate movements, providing no slippage occurs at the contact between plate and mantle. The convection currents might take hundreds of millions of years to complete a single loop. Some of these loops can be extremely large in the horizontal dimension and correspond to the dimensions of the associated plate. In the case of the Pacific plate, the loop would have to reach some 6,000 miles across.

Besides these large-scale features, small-scale convection cells might exist. Their horizontal dimensions would be comparable to a depth of about 410 miles, corresponding to the thickness of the upper mantle. Hot material rises from within the mantle and circulates horizontally near Earth's surface. There the top 30 miles or so cools to form the rigid plates, which carry the crust around. The plates complete the mantle convection by plunging back into Earth's interior. Thus, they are merely surface expressions of mantle convection.

Convection in the mantle would be expected to be strongly influenced by Earth's rotation. This is similar to the rotation's influence on air and ocean currents by the Coriolis effect, which bends poleward-flowing currents to the west and equatorward-flowing currents to the east (see chapter 6 for more on the Coriolis effect). Yet the rotation does not seem to affect the mantle. Even if convective flow occurred, it might not exist in neat circular cells. Instead, it might create eddy currents. The flow would thus become turbulent and extremely complex. Furthermore, the mantle is heated not only from below, but like the crust it is also heated from within by radioactive decay. This further complicates the development of convection cells

Figure 70 *The rifting of the African continent is occurring in the Red Sea, the Gulf of Aden, the Ethiopian rift valley, and the East African Rift.*

and causes distortion because the interior of the cells would no longer be passive. The interior would instead provide a significant portion of the heat as well.

Convection currents transport heat by the motion of mantle material, which in turn drives the plates. The mantle convection currents are believed to originate more than 410 miles below the surface. The deepest known earthquakes are detected at this level. Since plate motions trigger almost all large earthquakes, the energy they release must come from the forces that drive the plates. At the plate boundaries where one plate dives under another, the sinking slab meets great resistance to its motion at a depth of about 410 miles. This is the boundary between the upper and lower mantle, where the slabs tend to pile up.

However, sinking ocean crust has been known to breach this barrier and sink as much as 1,000 miles or more below the surface. Seismic images of mantle downwelling beneath the west coast of the Americas show a slab of subducting Pacific Ocean floor diving down to the very bottom of the mantle. Another slab of ancient ocean floor is sinking under the southern margin of Eurasia and is thought to be the floor of the Tethys, an ancient sea that separated India and Africa from Laurasia. Ocean slabs are also sinking into the mantle beneath Japan, eastern Siberia, and the Aleutian Islands.

If a slab should sink as far as the bottom of the lower mantle, it might provide the source material for mantle plumes called hot spots. If all oceanic plates were to sink to this level, a volume of rock equal to that of the entire upper mantle would be thrust into the lower mantle every 1 billion years. In order for the two mantle layers to maintain their distinct compositions, one floating on the other like oil on water, some form of return flow back to the upper mantle would be needed. Hot-spot plumes seem to fulfill this function.

The convection cells might also be responsible for the rising jets of magma that create chains of volcanoes, such as the Hawaiian Islands (Fig. 71). A strong mantle current possibly runs beneath the islands and disrupts the plume of ascending hot rock. Instead of rising vertically, the plume is sheared into discrete blobs of molten rock that climb like balloons in the wind. Each small plume created a line of volcanoes pointing in the direction of the movement of the underlying mantle. This might explain why the Hawaiian volcanoes do not line up exactly and why they erupt dissimilar lavas.

The asthenosphere is constantly losing material, which escapes from midocean ridges and adheres to the undersides of lithospheric plates. If the asthenosphere were not continuously fed new material from mantle plumes, the plates would grind to a complete halt. Earth would then become, in all respects, a dead planet because all geologic activity would cease.

Figure 71 *Photograph of the Hawaiian Island chain looking south, taken from the space shuttle. The main island, Hawaii, is in the upper portion of the photograph.*

(Photo courtesy NASA)

SEAFLOOR SPREADING

Seafloor spreading creates new lithosphere at spreading ridges on the ocean floor. It begins with hot rock rising from deeper portions of the mantle by convection currents. After reaching the underside of the lithosphere, the mantle rock spreads out laterally, dissipates heat near the surface, cools, and descends back into the deep interior of the Earth, where it receives more heat in a repeated cycle.

The constant pressure against the bottom of the lithosphere fractures the plate and weakens it. Convection currents flowing outward on either side of the fracture carry the separated parts of the lithosphere along with them, widening the gap. The rifting reduces the pressure in the underlying mantle, allowing mantle rocks to melt and rise through the fracture zone.

The molten rock passes through the lithosphere and forms magma chambers that supply molten rock for the generation of new lithosphere. Crustal material is sometimes introduced into the deep magma sources by subduction or off-scraping of a continental margin. The magma reservoirs resemble a mushroom up to 6 miles wide and 4 miles thick. The greater the supply of magma to the chambers, the higher they elevate the overlying spreading ridge.

As magma flows outward from the trough between ridge crests, it adds layers of basalt to both sides of the spreading ridge, creating new lithosphere. Some molten rock overflows onto the ocean floor in tremendous eruptions that generate additional oceanic crust. The continents ride passively on the lithospheric plates created at spreading ridges and destroyed at subduction zones. Therefore, the engine that drives the birth and evolution of rifts and, consequently, the breakup of continents and the formation of oceans ultimately originates in the mantle.

The plates grow thicker as they move away from a midocean spreading ridge as material from the asthenosphere adheres to the underside of the plates and transforms into new oceanic lithosphere. The continental plates vary in thickness from 25 miles in the young geologic provinces, where the heat flow is high, to 100 miles or more under the continental shields, where the heat flow is much lower. The shields are so thick they can actually scrape the bottom of the asthenosphere. The drag acts as an anchor to slow the motion of the plate.

The spreading ridges are the sites of frequent earthquakes and volcanic eruptions. The entire system acts as though it was a series of giant cracks in Earth's crust from which molten magma leaks out onto the ocean floor. Over much of its length, the ridge system is carved down the middle by a sharp break or rift that is the center of an intense heat flow. Magma oozing out at spreading ridges erupts basaltic lava through long fissures in the trough between ridge crests and along lateral faults. The faults usually occur at the boundary between lithospheric plates, where the oceanic crust pulls apart by the plate separation. Magma welling up along the entire length of the fissure forms large lava pools that harden to seal the fracture.

The spreading ridge system is not a continuous mountain chain. Instead, it is broken into small, straight sections called spreading centers (Fig. 72). The movement of new lithosphere generated at the spreading centers produces a series of fracture zones. These are long, narrow, linear regions up to 40 miles wide that consist of irregular ridges and valleys aligned in a stair-step shape. When lithospheric plates slide past each other as the seafloor spreads apart, they create transform faults ranging from a few miles to several hundred miles long. They are so named because they transform from active faults between spreading ridge axes to inactive fracture zones past the ridge axes. The transform faults partition the midocean ridge system into independent segments, each with their individual volcanic sources.

Figure 72 *Spreading centers on the ocean floor are separated by transform faults.*

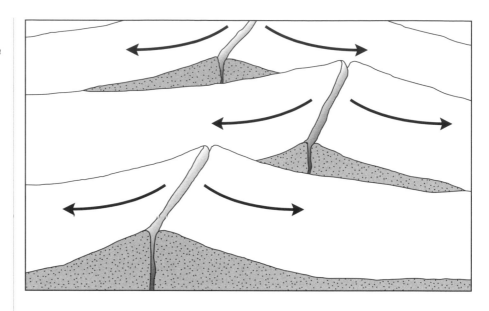

The transform faults of the Mid-Atlantic Ridge are offset laterally in a roughly east-west direction. The faults occur every 20 to 60 miles along the midocean ridge. The longer offsets consist of a deep trough joining the tips of two segments of the ridge. Other types of offsets up to 15 miles wide separate several spreading centers, which are each 20 to 30 miles long. The end of one spreading center often runs past the end of another. Sometimes the tips of the segments bend toward each other. Friction between segments produces strong shearing forces that wrench the ocean floor into steep canyons.

Transform faults appear to result from lateral strain, which is how rigid lithospheric plates are expected to react on the surface of a sphere. This activity is more intense in the Atlantic, where the spreading ridge system is steeper and more jagged than in the Pacific and Indian Oceans. Transform faults dissecting the Mid-Atlantic Ridge are generally more rugged than those of the East Pacific Rise. Moreover, fewer widely spaced transform faults exist along the East Pacific Rise, where the rate of seafloor spreading is five to 10 times faster than at the Mid-Atlantic Ridge. Therefore, the crust affected by transform faults is younger, hotter, and less rigid in the Pacific than in the Atlantic, giving the Pacific undersea terrain much less relief.

BASALTIC MAGMA

Most of Earth's surface above and below the sea is of volcanic origin. About 80 percent of all oceanic volcanism occurs along spreading ridges, where magma welling up from the mantle spews out onto the ocean floor. The seafloor on the crest of the midocean ridge consists of hard volcanic rock. The spreading crustal plates grow by the steady accretion of solidifying magma. The molten magma beneath the spreading ridges consists mostly of peridotite, an iron-magnesium silicate.

As the peridotite melts while rising through the lithosphere, a portion becomes highly fluid basalt. More than 1 square mile of new ocean crust, comprising about 5 cubic miles of basalt, forms throughout the world annually in this manner. However, sometimes gigantic flows erupt on the ocean floor with enough new basalt to pave the entire U.S. interstate highway system 10 times over.

Mantle material extruding onto the surface is black basalt, which is rich in silicates of iron and magnesium. Most of the world's nearly 600 active volcanoes are entirely or predominately basaltic. The magma from which basalt forms originated in a zone of partial melting in the upper mantle more than 60 miles below the surface. The semimolten rock at this depth is less dense than the surrounding mantle material and rises slowly toward the surface. As the magma ascends, the pressure decreases and more mantle material melts. Volatiles such as dissolved water and gases aid in making the magma flow easily.

Magma rising toward the surface fills shallow reservoirs or feeder pipes that are the immediate source of volcanic activity. The magma chambers closest to the surface exist under spreading ridges, where the oceanic crust is only 6 miles thick or less. Large magma chambers lie under fast spreading ridges where the lithosphere forms at a high rate, such as those in the Pacific. Narrow magma chambers lie under slow spreading ridges such as those in the Atlantic.

As the magma chamber swells with molten rock and begins to expand, the crest of the spreading ridge bulges upward due to the buoyant forces generated by the magma. The greater the supply of molten magma, the higher it elevates the overlying ridge segment. The magma rises in narrow plumes that balloon out along the spreading ridge, upwelling as a passive response to the release of pressure from plate divergence, somewhat like having the lid taken off a pressure cooker. Only the center of the plume is hot enough to rise all the way to the surface, however. If the entire plume erupted, it would build a massive volcano several miles high that would rival the tallest volcanoes in the solar system.

When the magma reaches the surface, it erupts a variety of gases, liquids, and solids. Volcanic gases mostly consist of steam, carbon dioxide, sulfur dioxide, and hydrochloric acid. The gases are dissolved in the magma and released as it rises toward the surface and pressures decrease. The composition of the magma determines its viscosity and type of eruption, whether quiet or explosive. If the magma is highly fluid and contains little dissolved gas when reaching the surface, it flows from a volcanic vent or fissure as basaltic lava. In such a case, the eruption is usually quite mild, as with Hawaiian Island volcanoes (Fig. 73).

The main types of lava formations associated with midocean ridges are sheet flows and pillow, or tube flows, which form pillow lavas (Fig. 74). Sheet flows are more prevalent in the active volcanic zone of fast spreading ridge segments such as those of the East Pacific Rise, where in some places the plates separate at a rate of 5 or more inches per year. These flows consist of flat slabs of basalt usually less than 1 foot thick. The basalt that forms sheet flows is more fluid than that responsible for pillow structures. Pillow lavas often occur at slow spreading ridges, such as at the Mid-Atlantic Ridge. There plates separate at a rate of only about 1 inch per year and the lava is much more viscous. The manufacture of new oceanic crust in this manner explains why some of the most intriguing terrain features lie on the bottom of the ocean

The Gorda Ridge, a deep-sea mountain range off the Northwest coast, forms where two oceanic plates abut one another. As the plates pull apart, they open up a crack that allows lava to rise from deep inside Earth. Oceanographers listening in on the Pacific Ocean through undersea microphones,

Figure 73 *An aa lava flow entering the sea from the January 21, 1960, eruption of Kilauea, Hawaii.*

(Photo by D. H. Richter, courtesy USGS)

planted on the seabed by the U.S. Navy to track submarines, heard the sound of a volcanic eruption along the Gorda Ridge. When researchers visited the ridge a few days later, they witnessed the actual birth of new ocean floor by seafloor spreading. While the eruption was in progress, they found a large pool of warmed water just above the ridge, whose summit at that point was 10,000 feet below sea level. After returning with a remote camera, they spotted fresh lava that had recently erupted. This was one of only a few times that scientists have caught seafloor spreading in the act.

Figure 74   Pillow lava on Knight Island, Alaska.

(Photo by F. H. Moffit, courtesy USGS)

THE CIRCUM-PACIFIC BELT

Deep-sea trenches, where the ocean floor disappears into Earth's interior, ring the Pacific. Lithospheric plates descend sheetlike into the mantle at subduction zones lying off continental margins and adjacent to island arcs. Plate subduction is responsible for the intense seismic activity that fringes the Pacific Ocean in a region known as the circum-Pacific belt, a chain of subduction zones flanking the Pacific basin.

Most earthquakes originate at plate boundaries (Fig. 75). Wide bands of earthquakes mark continental plate margins. Narrow bands of earthquakes

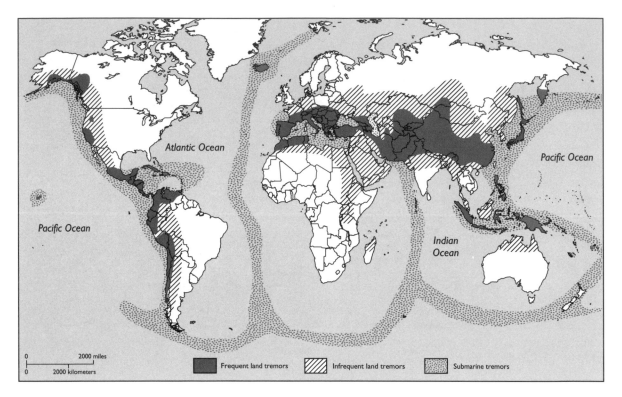

Frequent land tremors | Infrequent land tremors | Submarine tremors

Figure 75 Most earthquakes occur in broad zones associated with plate boundaries.

mark many major oceanic plate boundaries. The most powerful quakes are associated with plate subduction where one plate thrusts under another in deep subduction zones. The greatest amount of seismic energy occurs along the rim of the Pacific Ocean. In the western Pacific, the circum-Pacific belt encompasses volcanic island arcs that fringe the subduction zones, producing some of the largest earthquakes in the world.

The circum-Pacific belt is also known for its extensive volcanic activity. Subduction zone volcanoes form island arcs, mostly in the Pacific, and most volcanic mountain ranges on the continents. The circum-Pacific belt coincides with the Ring of Fire. The same tectonic forces that produce earthquakes are responsible for volcanic activity. This explains why the Pacific rim also contains the majority of the world's active volcanoes. The area of greatest seismicity is on the plate boundaries associated with deep trenches and volcanic island arcs, where an oceanic plate dives under a continental plate.

When starting from New Zealand, a land traversed by earthquake faults (Fig. 76), the circum-Pacific belt runs northward. It encompasses the islands of Tonga, Samoa, Fiji, the Loyalty Islands, the New Hebrides, and the Solomons. The belt then runs westward to embrace New Britain, New Guinea, and the Moluccas Islands. One segment continues westward over Indonesia. However, the principal arm travels northward to encompass the Philippines, where a large fault zone runs from one end of the islands to the other. The seismic belt continues on to Taiwan and the Japanese archipelago, which has been hard hit by major earthquakes. The January 17, 1995, Kobe earthquake of 7.2 magnitude killed more than 5,500 people and destroyed over $100 billion worth of property.

An inner belt runs parallel to the main belt and takes in the Marianas. This string of volcanic islands is characterized by a massive trench system in places more than 30,000 feet deep. The belt continues northward and follows the seismic arc across the top of the Pacific. It comprises the Kuril Islands (devastated by an 8.2 magnitude earthquake on October 4, 1994), the Kamchatka Peninsula, and the Aleutian Islands, which constantly rock and

Figure 76 The Wellington Fault, New Zealand.

(Photo courtesy USGS)

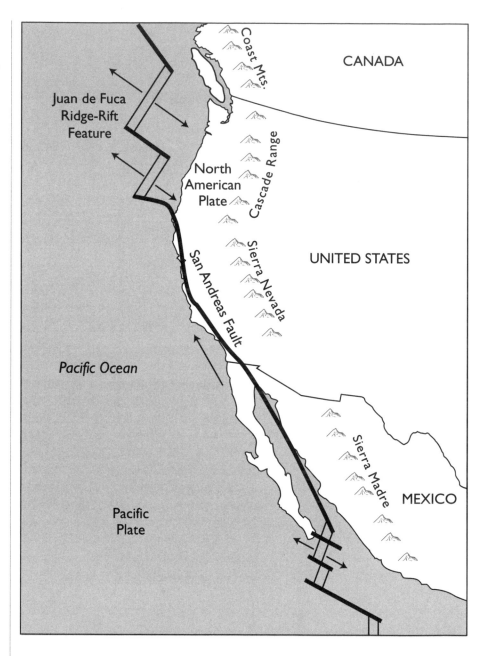

roll. The Aleutian Trench, the largest on Earth, is responsible for the many great earthquakes that strike Alaska. A 200-mile-long stretch called the Shumagin gap, which is accumulating huge stresses in the descending Pacific plate, is poised for a massive earthquake.

When crossing over to the eastern side of the Pacific basin, the seismic belt continues along the Cascadia subduction zone (Fig. 77). It extends along the coast from southern British Columbia to northern California. This belt has severely shaken the Pacific Northwest in the geologic past and is responsible for numerous powerful volcanoes. The Juan de Fuca and Gorda plates slipping under the North American plate generate the tectonic activity.

The San Andreas Fault marks the boundary between the North American and Pacific plates and rattles much of California (Fig. 78). The fault is a huge fracture zone 650 miles long and 20 miles deep. It runs northward from the Mexican border through southern California and plunges into the ocean at Cape Mendocino 100 miles south of the Oregon border. The fault represents the margin between the Pacific plate and the North American plate, which are moving relative to each other in a right lateral direction at a rate of nearly 2 inches per year. When the two plates snag and try to tear free of each other, earthquakes rumble across the region.

The mountain ranges of Mexico and Central America and the Andes Mountain regions of South America, especially in Chile and Peru, have been lashed by some of the largest and most destructive earthquakes. The 1960 Chilean earthquake of 9.5 magnitude, the largest in modern history, elevated a California-sized chunk of crust some 30 feet. In June 2001, Chile was struck again by a massive 8.1 magnitude earthquake that leveled cities and killed peo-

Figure 78 Collapsed buildings in the Marina District caused by the October 17, 1989, Loma Prieta earthquake, San Francisco, California.

(Photo by G. Plafker, courtesy USGS)

ple by the hundreds. In the last century, some two dozen earthquakes of 7.5 magnitude or greater have devastated the region.

An immense subduction zone lying just off the coast influences the whole western seaboard of South America. The lithospheric plate on which the South American continent rides forces the Nazca plate to buckle under, causing great tensions to build deep within the crust. The ocean floor's descent into the mantle is accompanied by devastating earthquakes as it grinds under the adjacent plate. While some rocks shove downward, others thrust to the surface to raise the Andes Mountains, making them the fastest growing mountain range on Earth. The resulting forces build great stresses into the entire region and, as the strain builds and the crust cracks, great earthquakes roll across the countryside.

THE DEEP-SEA TRENCHES

Figure 79   The subduction zones where lithospheric plates enter the mantle are marked by the deepest trenches in the world.

The creation of new lithosphere at midocean ridges is matched by the destruction of old lithosphere at subduction zones (Fig. 79). Deep trenches lying at the edges of continents or along volcanic island arcs mark the seaward boundaries of the subduction zones. As a lithospheric plate sinks into the mantle, the line of subduction creates a deep-sea trench. While the Pacific plate drifts toward the northwest, its leading edge dives into the mantle, form-

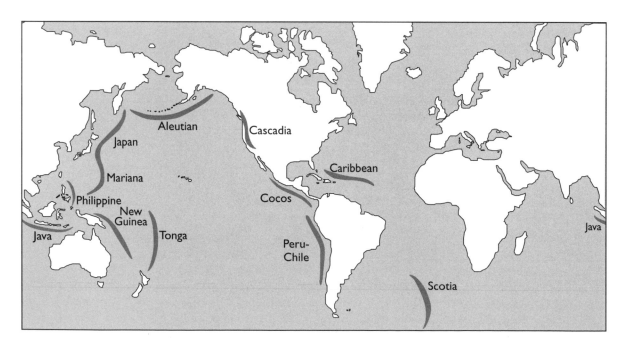

TABLE 10 THE WORLD'S OCEAN TRENCHES

Trench	Depth (miles)	Width (miles)	Length (miles)
Peru–Chile	5.0	62	3,700
Java	4.7	50	2,800
Aleutian	4.8	31	2,300
Middle America	4.2	25	1,700
Marianas	6.8	43	1,600
Kuril-Kamchatka	6.5	74	1,400
Puerto Rico	5.2	74	960
South Sandwich	5.2	56	900
Philippines	6.5	37	870
Tonga	6.7	34	870
Japan	5.2	62	500

ing the deepest trenches in the world (Table 10). The Mariana Trench in the western Pacific is the lowest point on Earth. It extends northward from the Island of Guam in the Mariana Islands and reaches a depth of nearly 7 miles below sea level.

Subduction zones, where cool, dense lithospheric plates dive into the mantle, are regions of low heat flow and high gravity (an area where the gravitational pull is strong relative to the average force of gravity on the surface). Conversely, because of their extensive volcanic activity, the associated island arcs are regions of high heat flow and low gravity. The deep-sea trenches are regions of intense volcanism, producing the most explosive volcanoes on Earth. Volcanic island arcs, which typically share similar curved shapes and similar volcanic origins, fringe the trenches. These island chains, for example the Aleutian Islands, are generally arc shaped because of the geometry of the ocean floor. The trenches outline an arc because this is the geometric figure formed when a plane cuts a sphere in the same manner that a rigid lithospheric plate subducts into the spherical mantle.

The trenches are also sites of almost continuous earthquake activity deep in the bowels of Earth, about 2 miles down. Plate subduction causes stresses to build into the descending lithosphere, producing deep-seated earthquakes that outline the boundaries of the plate. A band of shallow earthquakes clustered in a line running through Micronesia appears to mark the earliest stages in the birth of a subduction zone, indicated by the formation of a trench to the north and west of New Guinea in the western Pacific. Gravity in the area is lower than normal, as expected over a trench due to the sagging of the ocean floor. In addition, a bulge in the crust to the south of this area suggests that the edge of a slab of crust is beginning to dive into the mantle. The subduction process might not be operating fully for another 5 or 10 million years as the deep-sea trench nibbles away at the Pacific plate.

The seafloor south of New Zealand could also be experiencing the early stages of subduction in the process of creating a deep-sea trench. A geologic scar on the floor of the Pacific known as the Macquarie Ridge (Fig. 80) is still evolving as part of this process. The ridge is an undersea chain of mountains and troughs running south from New Zealand. It forms the boundary between the Australian and Pacific plates, which pass each other in opposite directions as the Australian plate moves northwest in relation to the Pacific plate. As a result of this action, in 1989, a massive earthquake of magnitude 8.2 struck the ridge.

As the Australian plate slides by the Pacific plate, ruptures occur along vertical faults where the plates slip past each other, creating large strike-slip

Figure 80 *The location of the Macquarie Ridge south of New Zealand.*

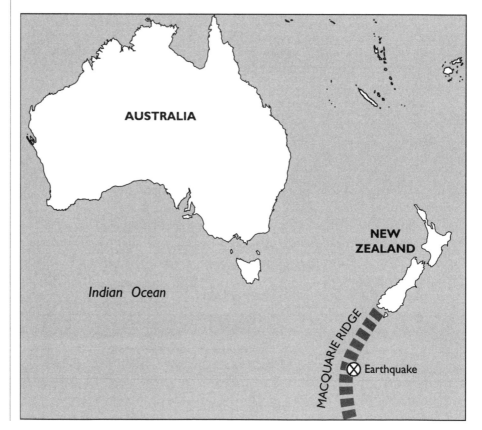

earthquakes. As they pass one another, the plates are also pressing together along dipping fault planes, creating smaller compressional earthquakes. This action suggests that subduction is just beginning along the Macquarie Ridge. However, the separate dipping faults that flank the area have not yet connected to form a single large fault plane, a necessary first step before subduction commences.

A plate extending away from its place of origin at a midocean spreading ridge becomes thicker and denser as additional material from the asthenosphere adheres to its underside in a process called underplating. The depth at which the oceanic crust sinks as it moves away from the midocean ridges varies with its age. Thus, the older the lithosphere the more basalt that underplates it, making the plate thicker, denser, and deeper.

Eventually, the plate becomes so dense it loses buoyancy and sinks into the mantle. The line of subduction creates a deep-sea trench at clearly defined subduction zones, where cool, dense lithospheric plates dive into the mantle. As the subducted portion of the plate dives into Earth's interior, the rest of the plate, which might carry a continent on its back, is pulled along with it like a freight train being hauled by a locomotive. Plate subduction is therefore the main driving force behind plate tectonics, and pull at subduction zones is favored over push at spreading ridges to move the continents around the surface of the globe.

New oceanic crust generated by seafloor spreading in the Atlantic and eastern Pacific is offset by the subduction of old ocean crust along the rim of the Pacific to make more room. Because the seafloor spreading rate is not always the same as the rate of subduction, associated midocean ridges often move laterally. Most subduction zones are in the western Pacific, which accounts for the fact that most oceanic crust is no older than 170 million years.

Subduction zones are the sites of almost continuous seismic activity. The band of earthquakes marks the boundaries of a sinking lithospheric plate (Fig. 81). As plates slide past each other along subduction zones, they create highly destructive earthquakes, such as those that have always plagued Japan, the Philippines, and other islands connected with subduction zones.

The subduction zones are also regions of intense volcanic activity, producing the most explosive volcanoes on the planet. Magma reaching the surface erupts on the ocean floor, creating new volcanic islands. Most volcanoes do not rise above sea level, however. Instead, they become isolated undersea volcanic structures called seamounts. The Pacific basin is more volcanically active and has a higher density of seamounts than the Atlantic or Indian basins. Subduction zone volcanoes are highly explosive because their magmas contain large quantities of volatiles and gases that escape violently when reaching the surface. The type of volcanic rock erupted in this manner is called andesite,

Figure 81 *A cross section of a descending lithospheric plate. Os denotes shallow earthquakes. Xs denotes deep-seated earthquakes.*

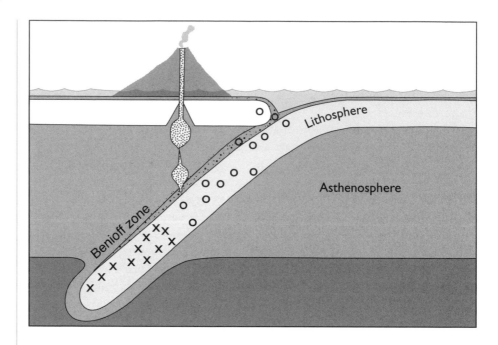

Figure 81 *A cross section of a descending lithospheric plate. Os denotes shallow earthquakes. Xs denotes deep-seated earthquakes.*

named for the Andes Mountains that form the spine of South America and well-known for their violent eruptions.

PLATE SUBDUCTION

The push that plates received from the expansion of the ocean floor at spreading ridges was once thought to be sufficient to force oceanic crust into the mantle at subduction zones. However, drag at the base of the plates can greatly resist plate motion. Therefore, an additional source of energy is needed to drive the plates. For this purpose, the force of gravity is called upon to provide the driving mechanism to overcome the resistance caused by plate drag. Therefore, the pull of a sinking slab of oceanic crust is the strongest force moving the plates around the surface of the globe.

An additional force that might help to overcome the resistance caused by plate drag is the pull the sinking plate receives by mantle convection currents. The magnitude of this force depends on the length of the subduction zone, the rate of subduction, and the amount of trench suction. With these forces in place, the plates could practically drive themselves without the aid of seafloor spreading. Therefore, the upwelling of magma at midocean spreading ridges might simply be a passive response to the plates being pulled apart by subduction.

As the rigid lithospheric plate carrying the oceanic crust descends into Earth's hot interior, it slowly breaks up and melts. Over a period of millions of years, it is absorbed into the general circulation of the mantle. When the plate dives into the interior, most of its trapped water goes down with it, becoming an important volatile in magma. The subducted plate also supplies molten magma for volcanoes, most of which ring the Pacific Ocean and recycle chemical elements to the planet.

The amount of subducted plate material is vast. When the Atlantic and Indian Oceans opened up and began forming new oceanic crust some 125 million years ago, an equal area of oceanic crust disappeared into the mantle. This meant that 5 billion cubic miles of crustal and lithospheric material was destroyed. At the present rate of subduction, the mantle will consume an area equal to the entire surface of the planet in just 160 million years.

The convergence of lithospheric plates forces the thinner, more dense oceanic plate under the thicker, more buoyant continental plate. When oceanic plates collide, the older and denser plate dives under the younger plate (Fig. 82). A deep-ocean trench marks the line of initial subduction. At first, the plate's angle of descent is low. However, it gradually steepens to about 45 degrees, with the rate of vertical descent (typically 2 to 3 inches per year) less than the rate of horizontal motion of the plate.

If continental crust moves into a subduction zone, its greater buoyancy prevents it from being dragged down into the trench. When two continental plates converge, the crust is scraped off the subducting plate and fastened onto the overriding plate, welding the two pieces of continental crust together. Meanwhile, the subducted lithospheric plate, now without its overlying crust, continues to dive into the mantle, squeezing the continental crusts together and forcing up mountain ranges.

In many subduction zones, such as the Lesser Antilles, sediments and their contained fluids are removed by offscraping and underplating in accretionary prisms. These are wedges of sediment that form on the overriding plate adjacent to the trench (Fig. 83). In other subduction zones such as the Mariana and Japan trenches, little or no sediment accretion occurs. Thus, subduction zones differ markedly from one another in the amount of sedimentary material removed at the accretionary prism. In most cases, at least some sediment and bound fluids appear to be subducted to deeper levels.

The underthrusting of continental crust by additional crustal material increases buoyancy and pushes up mountain ranges. Such a process occurred when India collided with Asia about 45 million years ago, raising the Himalayas. A strange series of east-west wrinkles in the ocean crust just south of India verifies that the Indian plate is still pushing northward, shrinking the Asian continent by as much as 3 inches a year. Further compression and deformation might take place beyond the line of collision, producing a high plateau

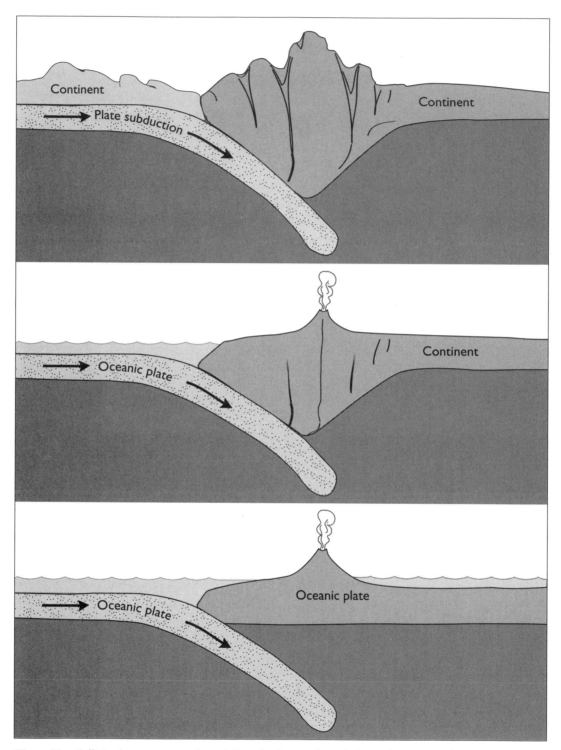

Figure 82 *Collision between two continental plates* (top), *a continental plate and an oceanic plate* (middle), *and two oceanic plates* (bottom).

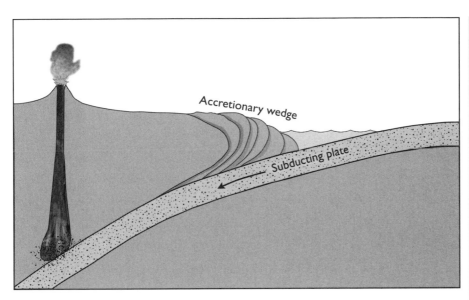

Figure 83 An accretionary wedge is formed by accumulating layers of a descending oceanic plate into Earth's crust.

with surface volcanoes similar to the Tibetan Plateau, the largest tableland in the world.

When continental and oceanic plates converge, the denser oceanic plate dives beneath the lighter continental plate and is forced farther downward. The sedimentary layers of both plates are squeezed, swelling the leading edge of the continental crust to create folded mountain belts such as the Appalachians. As the descending plate dives farther under the continent, it reaches depths where the temperatures are extremely high. The upper part of the plate melts to form magma that rises toward the surface to provide volcanoes with a new supply of molten rock.

After discussing ridges, trenches, and other major geologic structures on the seabed, the next chapter takes a look at volcanoes on the ocean floor.

5

SUBMARINE VOLCANOES
ERUPTIONS ON THE OCEAN FLOOR

This chapter examines volcanic activity on the ocean floor. An extraordinary number of volcanoes are hidden under the waves. More than 80 percent of the surface above and below the sea is of volcanic origin (Table 11). The vast majority of the volcanic activity that continually remakes the surface of Earth takes place at the bottom of the ocean, where most of the world's volcanoes are located. They accompany crustal movements at plate margins where lithospheric plates diverge at spreading ridges or converge at subduction zones.

Oceanic volcanoes also happen to be among the most explosive in the world. Whole islands have been known to disappear, while new ones pop up to take their places. Nearly all the world's islands started out as undersea volcanoes. In volcanic island building, successive eruptions pile up volcanic rocks until the volcano's peak finally breaks the surface of the sea. They originate as active undersea volcanoes that rise tens of thousands of feet off the ocean floor. Because these volcanoes arise from the very bottom of the ocean, they make the tallest mountains in the world.

TABLE 11   COMPARISON OF TYPES OF VOLCANISM

Characteristic	Subduction	Rift Zone	Hot Spot
Location	Deep ocean trenches	Midocean ridges	Interior of plates
Percent active volcanoes	80 percent	15 percent	5 percent
Topography	Mountains, island arcs	Submarine ridges	Mountains, geysers
Examples	Andies Mts., Japan Is.	Azores Is., Iceland	Hawaiian Is., Yellowstone
Heat source	Plate friction	Convection currents	Upwelling from core
Magma temperature	Low	High	Low
Magma viscosity	High	Low	Low
Volatile content	High	Low	Low
Silica content	High	Low	Low
Type of eruption	Explosive	Effusive	Both
Volcanic products	Pyroclasts	Lava	Both
Rock type	Rhyolite, Andesite	Basalt	Basalt
Type of cone	Composit	Cinder fissure	Cinder shield

THE RING OF FIRE

Most of the world's volcanoes are associated with crustal movements at the margins of lithospheric plates. The interaction of plates also creates earthquakes. An almost continuous Ring of Fire (Fig. 84) runs along the rim of the Pacific. It coincides with the circum-Pacific belt because the same tectonic processes that generate earthquakes also produce volcanoes. The greatest activity occurs on plate boundaries associated with deep trenches along volcanic island arcs and the margins of continents.

The Ring of Fire corresponds to a band of subduction zones surrounding the Pacific basin, which have devoured almost all the seafloor since the breakup of Pangaea. The oldest oceanic crust lies in a small patch off southeast Japan and is about 170 million years old. The rest of the ocean floor is on average only about 100 million years old. While subducting into the mantle, the oceanic crust melts to provide molten magma for volcanoes fringing the deep-sea trenches. This is why most of the 600 active volcanoes in the world lie in the Pacific Ocean, nearly half of which reside in the western Pacific region alone.

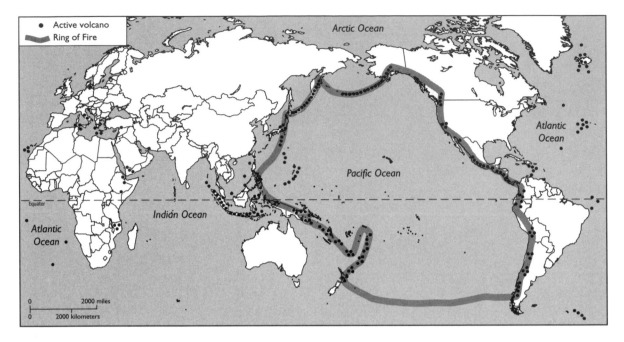

Figure 84 *The Ring of Fire is a band of subduction zones surrounding the Pacific Ocean.*

Subduction zone volcanism builds volcanic chains on the continents and island arcs in the ocean. At convergent plate boundaries, where one plate subducts under another, magma forms when the lighter constituents of the subducted oceanic crust melt. The upwelling magma creates island arcs, including Indonesia, the Philippines, Japan, the Kuril Islands, and the Aleutians—the longest, extending more than 3,000 miles from Alaska to Asia.

When beginning at the western tip of the Aleutian Islands off Alaska, the Ring of Fire runs along the Aleutian archipelago, a string of volcanic islands (Fig. 85) created by the subduction of the Pacific plate down the Aleutian Trench. The band of volcanoes turns south across the Cascade Range of British Columbia, Washington, Oregon, and northern California, associated with the subduction of the Juan de Fuca plate down the Cascadia subduction zone. The ring then runs across Baja California and southwest Mexico, where lie the volcanoes Parícutin (Fig. 86) and El Chichón. When Parícutin blew, it was one of the strangest eruptions of the 20th century because it originated in a farmer's cornfield. The eruption of El Chichón was perhaps the dirtiest of the 20th century in terms of ash cast high up into the atmosphere.

The volcano belt continues through western Central America, which has numerous active cones. For instance, the Nevado del Ruiz of Colombia produced destructive mudflows that killed 25,000 people in November 1985—one of the greatest volcanic disasters of the 20th century (Table 12). Some 20 volcanoes have been given special recognition by killing more than

Figure 85 *A crater and dome of Great Sitkin Volcano, Great Sitkin Island, Aleutian Islands, Alaska.*
(Photo by F. S. Simons, courtesy USGS)

Figure 86 *The cinder cone developed from the July 25, 1943, eruption of Parícutan Volcano, Michoacán, Mexico.*

(Photo by W. F. Foshag, courtesy USGS)

117

TABLE 12   MAJOR VOLCANIC DISASTERS OF THE 20TH CENTURY

Date	Volcano/Area	Fatalities	Remarks
1902	La Soufrière	15,000	
	Pelée (Martinique)	28,000	
	Santa María (Guatemala)	6,000	Deadliest volcano outbreak of 20th century, resulting in a total death toll of 35,000
1919	Keluit (Indonesia)	5,500	Deaths resulted from volcanic mudflows called lahars
1977	Nyiragongo (Zaire)	70	
1980	St. Helens (United States)	62	Worst volcanic disaster in the nation's recorded history
1982	Galunggung (Indonesia)	27	Relatively few deaths, but huge loss of property and suffering
1983	El Chichón (Mexico)	2,000	Worst volcanic disaster in the nation's recorded history
1985	Nevado del Ruiz (Colombia)	22,000	Worst volcanic disaster in the nation's recorded history
1986	Lake Nios (Cameroon)	20,000	
1991	Unzen (Japan)	37	Ongoing eruptions have forced 3,000 people form their homes
1991	Pinatubo (Philippines)	700	Largest eruption in nation's recorded history; remarkably few fatalities despite destruction
1993	Mayon (Philippines)	75	Eruptions produced deadly pyroclastic flows, 60,000 evacuated

1,000 people each since 1700. The Ring of Fire journeys along the course of the Andes Mountains on the western edge of South America, whose volcanoes' highly explosive nature results from the subduction of the Nazca plate down the Chilean Trench.

The volcanic band then turns toward Antarctica and the islands of New Zealand, New Guinea, and Indonesia, where the volcanoes Tambora and Krakatoa produced the greatest eruptions in modern history. These outbursts were initiated by the subduction of the Australian plate down the Java Trench. The band continues across the Philippines, where Mount Pinatubo created a massive eruption cloud in June 1991 that caused dramatic changes in climate, resulting from the subduction of the Pacific plate down the Philippine Trench. The Ring of Fire then runs across Japan, where the Fuji Volcano reigns majestically, finally ending on the Kamchatka Peninsula in northeast Asia, known for its powerful volcanoes.

Subduction zone volcanoes such as those in the western Pacific and Indonesia (Fig. 87 and Fig. 88) are among the most explosive in the world, often destroying entire islands when they erupt. One classic example is the near-total destruction of the Indonesian island of Krakatoa in 1883, which killed 36,000 people. The explosive nature of such volcanoes is due to abundant silica and volatiles in the magma, consisting of water and gases derived

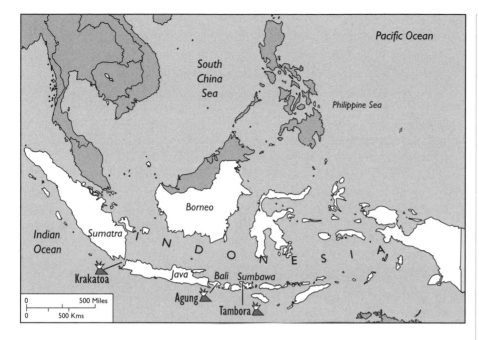

Figure 87 *Location of the great Indonesian volcanoes.*

from sediments on the ocean floor that were subducted into the mantle and melted. When the magma depressurizes as it reaches the surface, the volatiles explode and fracture the molten rock, destroying much of the volcano in the process.

On the continents, plate subduction forms long chains of powerful volcanoes. The Cascade Range in the Pacific Northwest is a belt of volcanoes associated with a subduction zone under the North American continent. The Andes Mountains of South America comprise a chain of volcanoes associated with a subduction zone under the South American continent. As the lithosphere plunges into the mantle, the tremendous heat melts portions of the descending plate and the adjacent lithospheric plate. The magma rising to the surface feeds rows of hungry volcanoes.

THE RISING MAGMA

Sediments washed off the continents and carried deep into the mantle in subduction zones melt to become the source of new molten magma for volcanoes fringing the deep-sea trenches. Some magma originates from the partial melting of subducted oceanic crust, with heat supplied by the shearing action at the top of the descending plate. Convective motions in the wedge of asthenosphere caught between the descending oceanic plate and the continental plate force material upward, where it melts under reduced pressures. The magma rises to the surface in giant blobs called diapirs. Upon reaching the underside of the lithosphere, the diapers burn holes through the crust as the molten rock melts on its journey upward.

As the diapirs rise toward the surface, they form magma bodies, which become the immediate source for new igneous activity. After reaching the surface, the magma erupts on the ocean floor to create new volcanic islands strung out on the ocean floor along with other volcanic activities (Fig. 89). Some of these volcanoes are extremely explosive because the magma contains large quantities of volatiles and gases that escape violently.

The rock type associated with subduction zone volcanoes is fine-grained gray andesite (Table 13). It contains abundant silica from deep-seated sources, possibly 70 miles below the surface. The rock derives its name from the Andes

Figure 88 *An active volcano on Andonara Island, Indonesia, leaves a 30-mile-long train of ash.*

(Photo courtesy NASA)

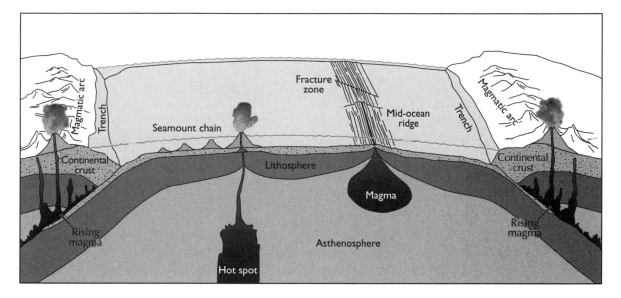

Mountains, whose volcanoes are highly explosive because of large amounts of volatiles in the magma. As the magma rises toward the surface, the pressure drops and volatiles escape with great force, shooting out of the volcano like pellets fired from a gigantic canon.

Figure 89 *Diapirs supply the magma for volcanoes and spreading ridges on the ocean floor.*

TABLE 13   CLASSIFICATION OF VOLCANIC ROCKS

Property	Basalt	Andesite	Rhyolite
Silica content	Lowest, about 50%, a basic rock	Intermediate, about 60%	Highest more than 65%, an acidic rock
Dark-mineral content	Highest	Intermediate	Lowest
Typical minerals	Feldspar Pyroxene Olivine Oxides	Feldspar Amphibole Pyroxine Mica	Feldspar Quartz Mica Amphibole
Density	Highest	Intermediate	Lowest
Melting point	Highest	Intermediate	Lowest
Molten rock viscosity at the surface	Lowest	Intermediate	Highest
Formation of lavas	Highest	Intermediate	Lowest
Formation of pyroclastics	Lowest	Intermediate	Highest

121

The mantle material that slowly extrudes onto the surface is black basalt, the most common volcanic rock. The ocean floor is paved with abundant basalt, and most volcanoes are entirely or predominately basaltic. The magma that forms basalt originates in a zone of partial melting in the upper mantle more than 60 miles below the surface. The semimolten rock at this depth is less dense and more buoyant than the surrounding mantle material and rises slowly toward the surface.

As the magma ascends, the pressure decreases, allowing more mantle material to melt. Volatiles, such as dissolved water and gases, make the magma flow easily. The mantle material below spreading ridges that create new oceanic crust consists mostly of peridotite, which is rich in silicates of iron and magnesium. As the peridotite melts as it progresses toward the surface, a portion becomes highly fluid basalt.

The magma's composition indicates its source materials and the depth within the mantle from which they originated. The degree of partial melting of mantle rocks, partial crystallization that enriches the melt with silica, and the assimilation of a variety of crustal rocks influence the composition of the magma. When the erupting magma rises toward the surface, it incorporates a variety of rock types along the way, which also changes its composition. The magma's composition determines its viscosity and the type of eruption that occurs.

If the magma is highly fluid and contains little dissolved gas, upon reaching the surface it produces basaltic lava, and the eruption is usually mild. If, however, the magma rising toward the surface contains a large quantity of dissolved gases, the eruption can be highly explosive and quite destructive. Water is possibly the single most important volatile in magma and affects the explosive nature of some volcanic eruptions by causing a rapid expansion of steam as the magma reaches the surface, where it creates new islands in the sea (Fig. 90).

ISLAND ARCS

Almost all volcanic activity is confined to the margins of lithospheric plates. Deep trenches at the edges of continents or along volcanic island arcs mark the seaward boundaries of subduction zones. At convergent plate boundaries, where one plate subducts under another, new magma forms when the lighter constituent of the subducted plate melts and rises to the surface. When the upwelling magma erupts on the ocean floor, it creates island arcs, which occur mostly in the Pacific.

The longest island arc is the Aleutian Islands, extending more than 3,000 miles from Alaska to Asia, where the Pacific plate ducks beneath the overriding North American plate. The Kurile Islands to the south form another long arc. The islands of Japan, the Philippines, Indonesia, New Hebrides, Tonga, and those from Timor to Sumatra also form island arcs. These island arcs are all similarly curved, have similar geologic compositions, and are associated with subduction zones. The curvature of the island arcs results from the curvature of Earth. Just as an arc forms when a plane cuts a sphere, so does an arc-shaped feature result when a rigid lithospheric plate subducts into the spherical mantle.

At deep-sea trenches, created during the subduction process, magma forms when oceanic crust that is thrust deep into the mantle melts. As the lithospheric plate carrying the oceanic crust descends farther into Earth's interior, it slowly breaks up and melts as well. Over a period of millions of years, it assimilates into the general circulation of the mantle, possibly descending as deep as the top of the core. Eventually, the magma rises to the surface in giant plumes, completing the loop in the convection of the mantle.

The subducted plate becomes the immediate source of molten magma for volcanic island arcs (Fig. 91). Behind each island arc is a marginal or a back-arc basin, a depression in the ocean crust due to the effects of plate subduction. Steep subduction zones such as the Mariana Trench in the western

Figure 90 *A submarine eruption of Myojin-sho Volcano in the Izu Islands, Japan.*

(Photo courtesy USGS)

Figure 91 *The formation of volcanic island arcs by the subduction of a lithospheric plate.*

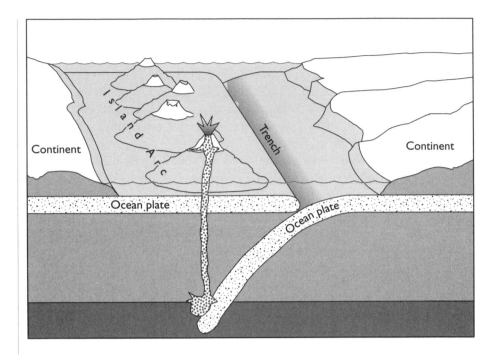

Pacific form back-arc basins, whereas shallow ones such as the Chilean Trench off the west coast of South America do not. A classic back-arc basin is the Sea of Japan (Fig. 92) between China and the Japanese archipelago, which is a combination of ruptured continental fragments. Gradually, the sea will close off entirely as the Japanese islands slam into Asia.

Back-arc basins are regions of high heat flow because they overlie relatively hot material brought up by convection currents behind the island arcs or by upwelling from deeper regions in the mantle. The trenches are regions of low heat flow because of the subduction of cool, dense lithospheric plates, while the adjacent island arcs are generally regions of high heat flow due to their high degree of volcanism.

GUYOTS AND SEAMOUNTS

Marine volcanoes associated with midocean ridges that rise above the sea become volcanic islands. Most of the world's islands began as undersea volcanoes. Successive volcanic eruptions pile up layers of volcanic rock until the peak finally breaks through the ocean surface. The volcanic ash also makes a rich soil. As the island cools, seeds carried by wind, sea, and animals rapidly turn the newly formed land into a lush tropical paradise. Life must still cope with the rumblings deep within Earth because the island could eventually be destroyed in a single huge convulsion.

Most volcanic islands end their lives quietly by the incessant pounding of the sea. Submarine volcanoes called guyots located in the Pacific once towered above the ocean. However, the constant wave action eroded them below the sea surface, leaving them as though the tops of the cones had been sawed off. The farther these volcanoes were conveyed from volcanically active regions, the older and flatter they became (Fig. 93). This suggests that the guy-

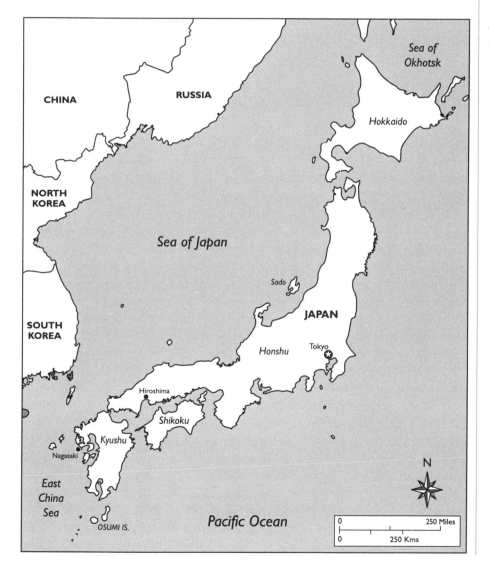

Figure 92   *The location of the Sea of Japan.*

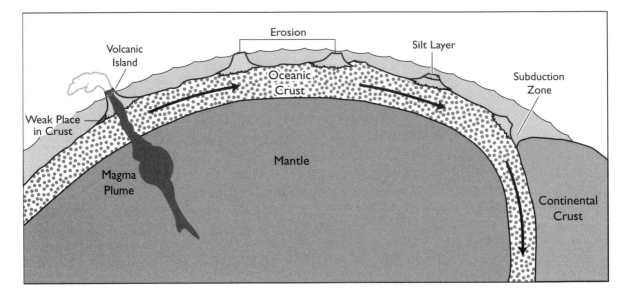

Figure 93 *Guyots were once active volcanoes that moved away from their magma source and have since disappeared beneath the sea.*

ots and the plates they rode on wandered across the ocean floor far from their places of origin. The islands appeared to have formed in assembly line fashion, each moving in succession away from a magma chamber lying beneath the ocean floor.

Beyond the oldest Hawaiian island, Kauai, the persistent pounding of the waves has eroded the volcanoes so that they now lie well below sea level. Coral living on the flattened tops of eroded volcanoes formed coral atolls, such as Midway Island, and shallow shoals. Atolls (Fig. 94) are rings of coral islands enclosing a central lagoon and consist of reefs up to several miles across. Many atolls formed on ancient volcanic cones that have subsided beneath the sea, with the rate of coral growth matching the rate of subsidence. Continuing in a northwestward direction is an associated chain of undersea volcanoes called the Emperor Seamounts (Fig. 95). These were presumably built by a single hot spot, although how such a plume could persist for more than 70 million years remains unexplained.

Most marine volcanoes never grow tall enough to rise above the sea and become islands. Instead, most remain as isolated undersea volcanoes called seamounts. Magma upwelling from the upper mantle at depths of more than 60 miles below the surface concentrates in narrow conduits that lead to the main feeder column. The magma erupts on the ocean floor, building elevated volcanic structures that form seamounts. These arc generally isolated and strung out in chains across the interior of a plate. Some seamounts are associated with extended fissures, along which magma wells up through a main conduit, piling successive lava flows on one another. The tallest seamounts rise more than 2.5 miles above the seafloor in the western Pacific near the Philippine Trench.

More than 10,000 seamounts rise up from the ocean floor. However, only a few, such as the Hawaiian Islands, manage to break the surface of the sea. The crust under the Pacific Ocean is more volcanically active than that of the Atlantic or Indian Oceans, providing a higher density of seamounts. The number of undersea volcanoes increases with advanced crustal age and increasing thickness. The average density of Pacific seamounts is 5 to 10 volcanoes per 5,000 square miles of ocean floor, by far outnumbering volcanoes on the continents.

Sometimes the summit of a seamount contains a crater, within which lava extrudes. If the crater exceeds 1 mile in diameter, it is called a caldera, whose depth below the crater rim is as much as 1,000 feet. Calderas form when the magma reservoir empties, creating a hollow chamber. Without support, the top of the volcanic cone collapses, forming a wide depression simi-

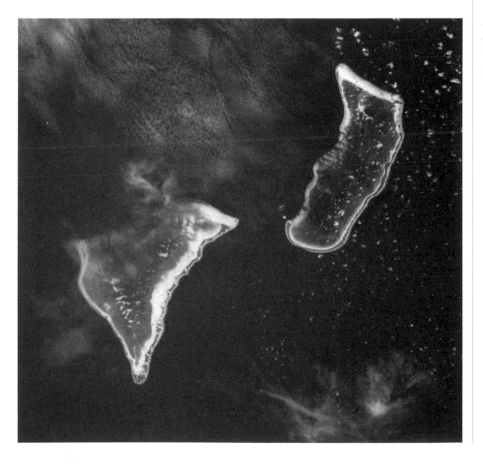

Figure 94 *Tarawa and Abaiang Atolls, Gilbert Islands.*

(Photo courtesy NASA)

Figure 95 The
Emperor Seamounts and
Hawaiian Islands in the
North Pacific represent
motions in the Pacific
plate over a volcanic hot
spot.

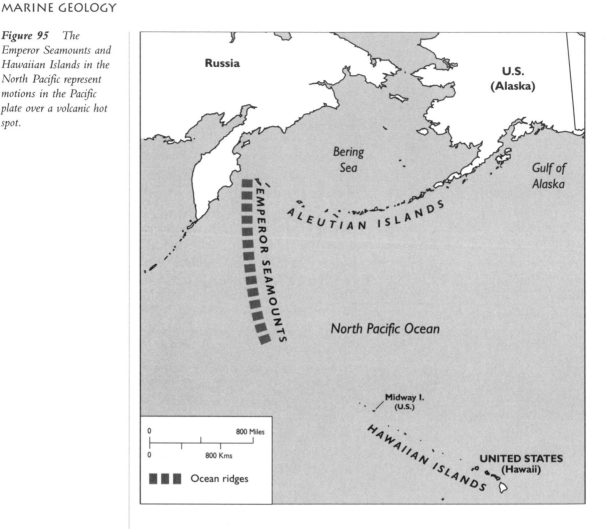

lar to calderas of Hawaiian volcanoes (Fig. 96). Feeder vents along the periphery of the undersea caldera supply fresh lava that fills the caldera, giving the volcano a flattop appearance. Other undersea volcanoes do not have a collapsed caldera. Instead, the summit contains several isolated volcanic peaks rising upward of 1,000 feet high.

RIFT VOLCANOES

More than three-quarters of oceanic volcanism occurs at midocean ridges, where basaltic magma wells up from the mantle and spews out onto the ocean floor in response to seafloor spreading. Deep-sea ridges called abyssal hills were developed by eruptions along midocean ridges and cover 60 to 70 percent of Earth's surface. Lithospheric plates subduct into the mantle like great slabs of rock and arise again in giant cylindrical plumes of hot magma at midocean ridges. A series of plumes miles apart feed separate segments of the spreading ridge.

At the crest of a midocean ridge, the ocean floor consists almost entirely of hard volcanic rock. Along much of its length, the ridge system is divided down the middle by a sharp break or rift that is the center of intense volcanic activity. The spreading ridges are the sites of frequent earthquakes and volcanic eruptions, as though the entire system were a series of giant cracks in the crust from which molten magma oozes out onto the ocean floor.

Volcanic eruptions associated with midocean rift systems are fissure eruptions, the most common type, and those that build typical conical volcanic structures. Fissure eruptions on the ocean floor occur at the boundaries between lithospheric plates where the brittle crust pulls apart by the process of seafloor spreading. Volcanoes formed on or near midocean ridges often develop into isolated peaks when they move away from the ridge axis as the seafloor spreads apart.

Figure 96   *A broad fountain pit in the cinder cone and large lava rivers draining from it, Halemaumau Volcano, Hawaiian Islands.*

(Photo by G. A. MacDonald, courtesy USGS)

During fissure eruptions, the magma oozes onto the ocean floor as lava that bleeds through fissures in the trough between ridge crests and along lateral faults. The faults usually occur at the boundary between lithospheric plates, where the oceanic crust splits apart by the separating plates. Magma welling up along the entire length of the fissure forms large lava pools, similar to those of broad shield volcanoes.

The lava formations that erupt on the midocean ridges are sheet flows and pillow, or tube, flows. Sheet flows are more prevalent in the active volcanic zone of fast spreading ridge segments, such as those of the East Pacific Rise. They consist of flat slabs of basalt usually less than 8 inches thick. The lava that forms sheet flows is much more fluid than that responsible for pillow formations. Pillow lavas appear as though basalt were squeezed out onto the ocean floor. They are mostly found in slowly spreading ridges such as the Mid Atlantic Ridge, where the lava is much more viscous. The surface of the pillows often has corrugations or small ridges pointing in the direction of flow. The pillow lavas typically form small, elongated hills descending downslope from the crest of the ridge.

Seamounts associated with midocean ridges that grow tall enough to break through the surface of the ocean become volcanic islands. The Galápagos Islands (Fig. 97) west of Ecuador are volcanic islands associated with the East Pacific Rise. The volcanic islands associated with the Mid–Atlantic Ridge include Iceland, the Azores, the Canary and Cape Verde Islands off West Africa, Ascension Island, and Tristan de Cunha.

The volcanic islands in the middle of the North Atlantic that comprise the Azores were created by a mantle plume or hot spot that once lay beneath Newfoundland, which then drifted westward as the ocean floor spread apart at the Mid–Atlantic Ridge. The Sts. Peter and Paul Islands in the mid–Atlantic north of the equator are not volcanic in origin. Instead, they are fragments of the upper mantle uplifted near the intersection of the St. Paul transform fault and the Mid–Atlantic Ridge.

Iceland is a broad volcanic plateau of the Mid–Atlantic Ridge that rose above the sea about 16 million years ago when the ridge assumed its present position. It is the most striking example of rift zone hot-spot volcanism. The magma plume underlying the island extends to the very base of the mantle some 1,800 miles down. What makes the island unique is that it straddles the Mid–Atlantic Ridge, where the two plates of the Atlantic basin and adjacent continents pull apart. Along the ridge, the abnormally elevated topography extends in either direction about 900 miles, with more than one-third of the plateau lying above sea level. South of Iceland, the broad plateau tapers off to form the typical Mid–Atlantic Ridge.

A steep-sided, V-shaped valley runs northward across the entire length of the island and is one of the few expressions of a midocean rift on land.

Figure 97   *The Galápagos Islands west of Ecuador.*

Numerous volcanoes flank the rift, making Iceland one of the most volcanically active places on Earth (Fig. 98). The powerful upwelling currents deep within the mantle produce glacier-covered volcanic peaks up to 1 mile high. In 1918, an eruption under a glacier unleashed a flood of meltwater 20 times greater than the flow of the Amazon, the world's largest river. Iceland experienced another under-ice eruption in 1996, when massive floods from gushing meltwaters and icebergs dashed 20 miles to the seacoast. Icelanders have known these glacial bursts called jokulhlaups since the 12th century.

On other parts of the midocean ridge, volcanic activity is quite prevalent. Perhaps as many as 20 major, deep underwater eruptions occur each year. Volcanoes formed on or near the midocean ridges often develop into isolated peaks as they move outward from the ridge axis during seafloor spreading. The ocean floor thickens as it moves away from the spreading ridge axis. This thickening of the seafloor influences a volcano's height because a thicker crust can support a greater mass. The ocean crust also bends like a rubber mat under

Figure 98   Seawater is sprayed onto the lava flow from the outer harbor of Vestmannaeyjar, Iceland, from the May 1973 eruption on Heimaey.

(Photo Courtesy USGS)

the massive weight of a seamount. For instance, the crust beneath Hawaii bulges in a downward concave shape as much as 6 miles.

A volcano formed at a midocean ridge cannot increase its mass unless it continues to be supplied with magma after it leaves the vicinity of the ridge. Sometimes a volcano formed on or near a midocean ridge develops into an island, only to have its source of magma cut off. Then erosion begins to wear it down until it finally sinks beneath the sea.

HOT-SPOT VOLCANOES

About 100 small regions of isolated volcanic activity known as hot-spot volcanoes exist in various parts of the world (Fig. 99). The hot spots provide a pipeline for transporting heat from the planet's core to the surface. The plumes do not rise through the mantle as a continuous stream, however. They instead bubble up as separate giant blobs of hot rock. When the bubbles reach the ocean floor at the top of the mantle, they create a succession of oceanic islands.

The ascending mantle plumes can lift an entire region. For example, a 3,000-mile-wide section of the South Pacific floor has risen where several hot spots have erupted to form the Polynesian Islands. Similar swells occur under the Hawaiian chain in the North Pacific, Iceland in the North Atlantic, and the Kerguelen Islands in the southern Indian Ocean. The most active modern hot spots lie beneath the big island of Hawaii and Réunion Island to the east of Madagascar.

Unlike most other active volcanoes, those created by hot spots are rarely situated at plate boundaries. Instead, they reside deep in the interiors of lithospheric plates (Fig. 100). Hot-spot volcanoes are notable for their geologic isolation far from normal centers of volcanic and earthquake activity. Lavas of hot-spot volcanoes differ markedly from those of subduction zones and rifts. The distinctive composition of hot-spot magmas suggests that their source is outside the general circulation of the mantle.

The lavas comprise basalts that contain larger amounts of alkali minerals such as sodium and potassium, indicating their source material is not connected with plate margins. Instead, the hot spots are supplied from deep

Figure 99 The world's hot spots, where mantle plumes rise to the surface.

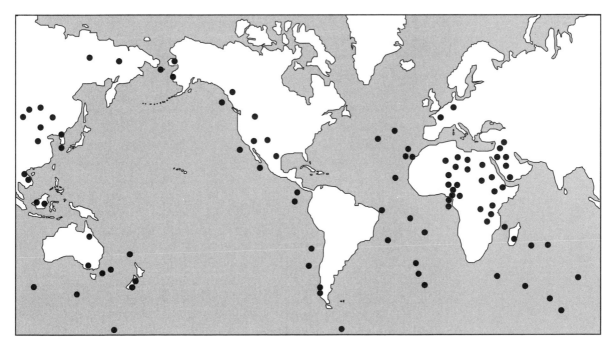

within the mantle, possibly near the top of the core. Hot-spot plumes might also arise from stagnant regions in the center of convection cells or from below the region in the mantle stirred by convection currents.

As plumes of mantle material flow upward into the asthenosphere, the portion rich in volatiles rises toward the surface to feed hot-spot volcanoes. The plumes exist in a range of sizes that might indicate the depth of their source material. They are not necessarily continuous flows of mantle material but might consist of molten rock rising in giant blobs or diapirs. If the upwelling plumes stopped feeding the asthenosphere with a continuing flow of mantle material, the plates would grind to a complete halt.

The typical life span of a plume is a few hundred million years. Sometimes a hot spot fades away and a new one forms in its place. The position of a hot spot changes slightly as it sways in the convective currents of the mantle. As a result, the hot-spot tracks on the surface might not always be linear.

However, compared with the motion of the plates, the mantle plumes are virtually stationary. Because the motion of the hot spots is so slight, they provide a reference point for determining the direction and rate of plate travel.

The passage of a plate over a hot spot often results in a trail of volcanic features whose linear trend reveals the direction of plate motion. This produces volcanic structures aligned in a direction that is oblique to the adjacent midocean ridge system rather than parallel to it as with rift volcanoes. The hot-spot track might be a continuous volcanic ridge or a chain of volcanic islands and seamounts that rise high above the surrounding seafloor. The hot-spot track can also weaken the crust, cutting through the lithosphere like a geologic blowtorch.

The most prominent and easily recognizable hot spot created the Hawaiian Islands (Fig. 101), the largest of their kind in the world. The youngest and most volcanically active island is Hawaii at the southeast end of the chain. One of the most volcanically active places on Earth is the erupting Kilauea Volcano on Hawaii (Fig. 102). Every day, several hundred thousand cubic yards of molten rock gush from a rift zone along its flanks. When the lava has run its course down the mountainside, it flows into the ocean, adding acres of new land to the island. The source of these fiery conditions is a mantle plume of hot rock burning through the Pacific plate from deep inside Earth. The hot rock has fueled the five volcanoes that built the Big Island of Hawaii.

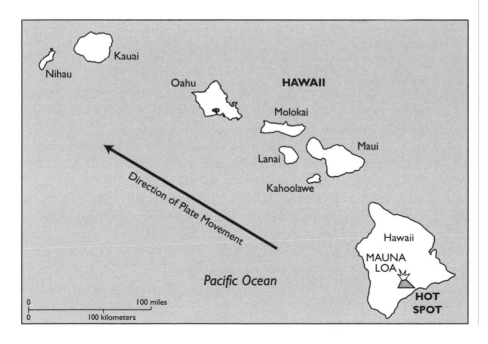

Figure 101 *The Hawaiian Islands formed by the drifting of the Pacific plate over a hot spot.*

Figure 102 *A lava flow entering the sea from the March 28, 1955, eruption of Kilauea Volcano, Hawaii.*

(Photo by G. A. MacDonald, courtesy USGS)

The oldest volcano, Kohala, on the northernmost part of the island, last erupted about 60,000 years ago. Today it is worn and eroded, with its northeastern flank deeply incised by spectacular valleys and gorges. Just to the south stands Mauna Kea, which happens to be the tallest mountain on Earth, rising more than 6 miles from the ocean floor. Southwest of Mauna Kea lies Hualalai Volcano, which last erupted in 1801 and is still actively, poised for another outburst. Southeast of Hualalai is Mauna Loa, the world's largest shield volcano. It consists of some 24,000 cubic miles of lava that built up flow upon flow into a huge, gently sloping mound, making it the most voluminous mountain on Earth. The youngest volcano, Kilauea, emerges from the side of Mauna Loa. Lava has erupted continuously since the early 1980s from a rift zone on Kilauea, which over time could greatly outgrow its host volcano.

Some 20 miles south of Hawaii is a submerged volcano called Loihi, which rises about 8,000 feet above the ocean floor but is still 3,000 feet below the sea surface. Perhaps in another 50,000 years it will rise above sea and take its place as the newest member of the Hawaiian chain. The rest of the Hawaiian Islands are progressively older, with extinct volcanoes trailing off to the northwest.

The entire Hawaiian chain apparently formed from a source of magma from the deepest part of the mantle over which the Pacific plate has passed in a northwesterly direction. The volcanic islands slowly popped out onto the ocean floor conveyor-belt fashion, with the oldest trailing off to the northwest farthest away from the hot spot. Similar chains of volcanic islands lie in the

Pacific and trend in the same general southeast-to-northwest direction as the Hawaiian Islands (Fig. 103). This indicates that the Pacific plate is moving off in the direction defined by the line of volcanoes. Lying parallel to the Hawaiian Islands are the Austral and Tuamotu ridges. The islands and seamounts were formed by the northwestward motion of the Pacific plate over a volcanic hot spot.

The plate did not always travel in this direction, however. Some 43 million years ago, it turned and followed a more northerly heading. The course change possibly resulted from a collision between the Indian and Asian plates and appears as a distinct bend in the hot-spot tracks. A sharp bend in the long Mendocino Fracture Zone jutting out of northern California confirms that the Pacific plate abruptly changed direction at the same time as the India-Asia plate

Figure 103 *The linearity of volcanic islands on the Pacific plate in the direction of movement.*

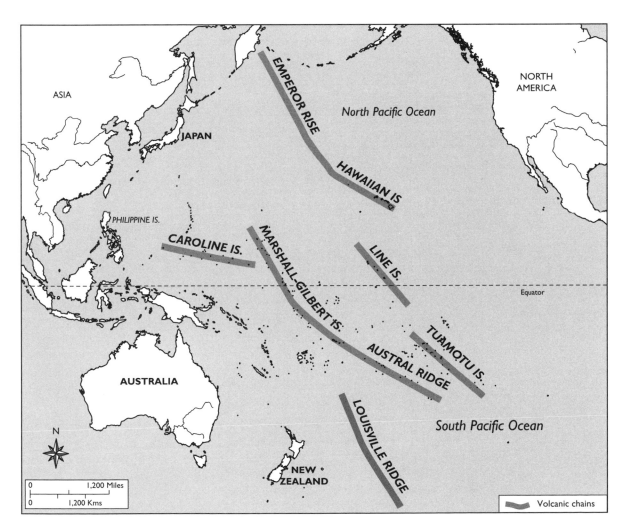

convergence. The timing is also coincident with the collision of the North American and Pacific plates. From these observations, geologists conclude that hot spots are generally a reliable means for determining plate activity.

The Bermuda Rise in the western Atlantic appears to contradict this rule. It is oriented in a roughly northeast direction, parallel to the continental margin off the eastern United States. The Bermuda Rise is nearly 1,000 miles long and rises some 3,000 feet above the surrounding seafloor. The last of its volcanoes ceased erupting about 25 million years ago. A weak hot spot unable to burn a hole through the North American plate was apparently forced to take advantage of previous structures on the ocean floor acting as conduits, which explains why the volcanoes trend nearly at right angles to the motion of the plate.

The Bowie seamount is the youngest in a line of submerged volcanoes running toward the northwest off the west coast of Canada. A mantle plume feeds it more than 400 miles below the ocean floor and is nearly 100 miles in diameter. Rather than lying directly beneath the seamount, as plumes usually do, this one lies about 100 miles east of the volcano. The plume might have taken a tilted path upward, or the seamount might have somehow moved with respect to the hot spot's position.

If a midocean ridge passes over a hot spot, the plume augments the flow of molten rock welling up from the asthenosphere to form new crust. The crust is therefore thicker over the hot spot than along the rest of the ridge, resulting in a plateau rising above the surrounding seafloor. The Ninetyeast Ridge, named for its location at 90 degrees east longitude, is a succession of volcanic outcrops that runs 3,000 miles south of the Bay of Bengal. It formed when the Indian plate passed over a hot spot on its way to Asia about 120 million years ago, creating an immense lava field on India known as the Rajmahal Traps.

The movement of the continents was more rapid than today, with perhaps the most vigorous plate tectonics the world has ever known. This activity resulted in many flood basalt eruptions (Fig. 104). About 120 million years ago, an extraordinary burst of submarine volcanism struck the Pacific basin, releasing vast amounts of lava onto the ocean floor. The volcanic spasm is evidenced by a collection of massive undersea lava plateaus that formed almost simultaneously. The largest of these plateaus is the Ontong Java, northeast of Australia. At roughly two-thirds the size of the Australian continent, it contains at least 9 million cubic miles of basalt, enough to bury the entire United States under 15 feet of lava.

About 65 million years ago, a giant rift opened up along the west side of India. Huge volumes of molten lava poured onto the surface, forming the Deccan Traps flood basalts. The rift separated the Seychelles Bank from the mainland, leaving behind the Seychelles Islands. They were followed 40 million years ago by the Kerguelen Islands as India continued to trek northward toward southern Asia.

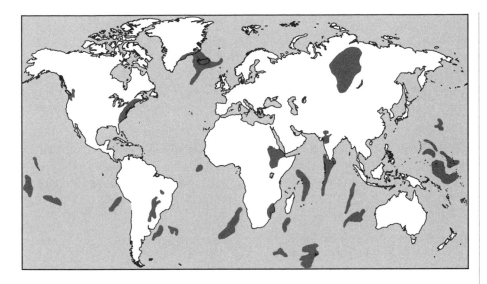

Figure 104 Areas of major flood basalt volcanism.

The Kerguelen plateau is the world's largest submerged platform. Approximately 50 million years ago, a huge submerged plateau in the Indian Ocean separated into two platforms that now sit about 1,200 miles apart. The plateau grew from the ocean floor more than 90 million years ago when a series of volcanic eruptions poured out voluminous amounts of molten basalt onto the Antarctic plate as the continent separated from Australia.

During the next several million years, a long rift sliced through the plate and cut off its northern section, which latched onto the Indian plate and started on a long journey northward. Meanwhile, the southern half of the plate continued to move southward. Half of the original platform, called the Broken Ridge, currently lies off the west coast of Australia. The other half, the Kerguelen plateau, sits north of Antarctica. The Exmouth plateau is a submerged feature that sits on a sunken piece of the Australian continent, which itself was attached to India when all continents were assembled into Pangaea.

VOLCANIC ACTIVITY

Volcanoes take many shapes and sizes, depending on the composition of the erupted magma. The four main types of volcanoes are cinder cones, composite volcanoes, shield volcanoes, and lava domes. Cinder cones are the simplest volcanic structures. They are built from particles and congealed lava ejected from a single vent. The 100-mile-long St. Lawrence Island in the North Bering Sea was built up by a number of cinder cones (Fig. 105). Explosive eruptions form short, steep slopes usually less than 1,000 feet high. Cinder cones build upward and

Figure 105 *St. Lawrence Island in the Bering Sea, showing cinder cones at the northwest end of Kookooligit Mountains.*

(Photo by H. B. Allen, courtsey USGS)

outward by accumulating layers of pumice, ash, and other volcanic debris falling back onto the volcano's flanks. The general order of events is eruption, followed by formation of cone and crater, and then lava flow.

Composite volcanoes are constructed from cinder and lava cemented into tall mountains rising several thousand feet. They are generally steep sided and comprise symmetrical lava flows, volcanic ash, cinders, and blocks. The crater at the summit contains a central vent or a cluster of vents. During eruption, the hardened plug in the volcano's throat breaks apart by the buildup of pressure from trapped gases below. The pent-up pressures shoot molten rock and fragments high into the air.

The fragments then fall back onto the volcano's flanks as cinder and ash. Layers of lava from milder eruptions reinforce the fragments, forming cones with a steep summit and steeply sloping flanks. Lava also flows through breaks in the crater wall or from fissures on the flanks of the cone, continually building it upward. As a result, composite volcanoes are the tallest cones in the world and often end in catastrophic collapse, thus preventing them from becoming the highest mountains.

Shield volcanoes produce the broadest and largest cones. The slope along their flanks generally rises only a few degrees and no more than 10 degrees near the summit. They erupt almost entirely basaltic lava from a central vent. Highly fluid molten rock oozes from the vent or violently squirts out, forming fiery fountains of lava (Fig. 106).

As the lava builds in the center, it flows outward in all directions, forming a structure similar to an inverted dinner plate. The lava spreads out and

Figure 106 A high lava fountain from an eruption of Kilauea, Hawaii.

(Photo by D. H. Richter, courtesy USGS)

covers large areas, as much as 1,000 square miles. If the lava is too viscous or heavy to flow very far, causing it to pile up around the vent, it forms a lava dome, which grows by expansion from within. Lava domes commonly occur in a piggyback fashion within the craters of large composite volcanoes.

Volcanoes erupt a variety of rock types. These range from rhyolite with a high silica content to basalt with a low silica and high iron-magnesium content. Basalt is the heaviest volcanic rock and the most common igneous rock type produced by the extrusion of magma onto the surface. Pumice is the lightest volcanic rock and can actually float on water. For instance, during the August 27, 1883, eruption of Krakatoa, floating pumice several feet thick posed a hazard to shipping in the area.

Tephra, from the Greek, meaning "ash," includes all solid particles ejected into the atmosphere from volcanic eruptions. Tephra includes an assortment of fragments from large blocks to dust-sized material. It originates from molten rock containing dissolved gases that rises through a conduit and suddenly separates into liquid and bubbles when nearing the surface. As the pressure decreases, the bubbles grow larger. If this event occurs near the orifice, a mass of froth spills out and flows down the sides of the volcano, forming pumice. If the reaction occurs deep down in the throat, the bubbles expand explosively and burst the surrounding liquid, which fractures the magma into fragments. The fragments are then driven upward by the force of the rapid expansion and hurled high above the volcano.

Cinder and ash supported by hot gases originating from a lateral blast of volcanic material is called nuée ardente, French for "glowing cloud." The cloud of ash and pyroclastics flows streamlike near the ground and might follow existing river valleys for tens of miles at speeds upward of 100 miles per hour. The best-known example was the 1902 eruption of Mt. Pelée, Martinique. It produced a 100-mile-per-hour ash flow that within minutes killed 30,000 people at Saint-Pierre. When the tephra cools and solidifies, it forms deposits called ash-flow tuffs that can cover an area up to 1,000 square miles or more.

Lava is molten magma that reaches the throat of a volcano or fissure vent and flows freely onto the surface. The magma that produces lava is much less viscous than that which produces tephra. This allows volatiles and gases to escape with comparative ease and thus gives rise to much quieter and milder eruptions. The outpourings of lava have Hawaiian names, Aa (pronounced, AH-ah), which is the sound of pain when walking over them barefooted, is blocky lava that forms when viscous, subfluid lavas press forward, carrying along a thick and brittle crust. As the lava flows, it stresses the overriding crust, breaking it into rough, jagged blocks. These are pushed ahead of, or dragged along with, the flow in a disorganized mass.

Pahoehoe, (pronounced pah-HOE-ay-hoe-ay), which means satinlike, are ropy lavas (Fig. 107) that are highly fluid basalt flows produced when the

Figure 107   *Ropy lava surface of pahoehoe near Surprise Cave, the Craters of the Moon National Monument, Idaho.*

(Photo by H. T. Stearns, courtesy USGS)

surface of the flow congeals, forming a thin, plastic skin. The melt beneath continues to flow, molding and remolding the skin into billowing or ropy-looking surfaces. When the lavas eventually solidify, the skin retains the appearance of the flow pressures exerted on it from below. If a stream of lava forms a crust and hardens on the surface and the underlying magma continues to flow away, it creates a long cavern or tunnel called a lava tube. Long caverns beneath the surface of a lava flow are created by the withdrawal of lava as the surface hardens. In exceptional cases, they can extend up to several miles within a lava flow.

The output of lava and pyroclastics for a single volcanic eruption varies from a few cubic yards to as much as 5 cubic miles. Rift volcanoes generate about 2.5 billion cubic yards per year, mainly submarine flows of basalt. Subduction zone volcanoes produce about 1 billion cubic yards of pyroclastic volcanic material per year. Volcanoes over hot spots produce about 500 million cubic yards per year, mostly pyroclastics and lava flows on the continents and basalt flows in the oceans.

After discussing marine volcanism, the next chapter explains the role ocean currents play on Earth.

6

ABYSSAL CURRENTS

OCEAN CIRCULATION

This chapter examines ocean currents, tides, and waves. The ocean is constantly in motion, distributing water and heat to all corners of the globe (Fig. 108). In effect, the ocean acts as a huge circulating machine that makes Earth's climate equitable. Ocean currents follow well-defined courses. The ocean transports tremendous quantities of seawater, serving as a global conveyor belt over the planet. Abyssal storms stir the deep-ocean floor, shifting sediments on the seabed.

El Niño currents, caused by a great sloshing of seawater in the Pacific basin, generate unusual weather patterns throughout the world. Waves and tides are constantly changing and rearranging the shoreline. Tsunamis produced by undersea earthquakes and coastal volcanic eruptions are among the most destructive waves, inflicting death and destruction to many seacoast inhabitants.

RIVERS IN THE ABYSS

Currents in the upper regions of the ocean (Fig. 109) are driven by the winds, which impart their momentum to the ocean's surface. The currents do not

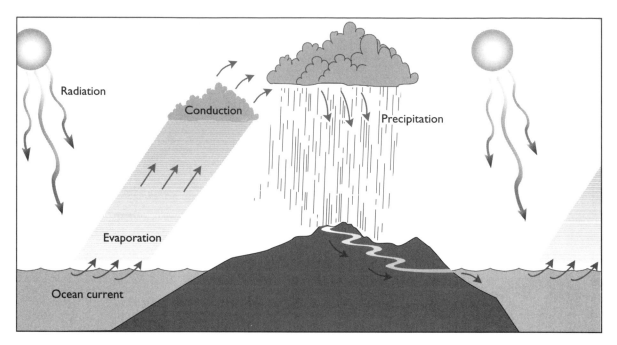

Figure 108 *Heat flow*
between the ocean and
atmosphere is responsible
for distributing the ocean's
heat around the world.

flow in the wind direction. Instead, they are deflected by the Coriolis effect (Fig. 110) to the right, or northwest, in the Northern Hemisphere and to the left, or southwest, in the Southern Hemisphere. This is because a point on the surface moves faster at the equator than if it were near the poles because it is farther away from the axis of rotation and must travel a greater distance over time. Currents moving toward the poles experience the surface slowing down, which deflects them to the east. Currents moving toward the equator experience the surface speeding up, which deflects them toward the west.

The currents acquire warm water from the tropics and distribute it to the higher latitudes. They then return to the tropics with cold water. This exchange moderates the temperatures of coastal regions, making countries such as Japan and northern Europe warmer than they otherwise would be for their latitudes. The Gulf Stream snakes 13,000 miles clockwise around the North Atlantic basin, transporting warm tropical water to the northern regions.

The North Atlantic acts as a tremendous heat pump that periodically warms and cools the atmosphere over the course of decades, sometimes reinforcing and sometimes counteracting greenhouse warming. The counterpart of the Gulf Stream in the North Pacific is another strong flow called the Japan current. This current bears warm water from the tropics, sweeps northward against Japan, crosses the upper Pacific, and turns southward to warm the western coast of North America. The major current in the South Pacific is the

Humboldt or Peru current, which flows northward along the west coast of South America.

Like huge undersea tornadoes, eddies or gyres of swirling warm and cold water accompany the ocean currents. Many eddies are enormous—measuring as much as 100 miles or more across and reaching depths of 3 miles. Most eddies, however, are less than 50 miles across. Some, including those in the Arctic Ocean off Alaska, are only 10 miles wide. Like giant eggbeaters, these small eddies play an important role in mixing the oceans.

The eddies appear to be pinched-off sections of the main ocean currents. As with high pressure systems in the atmosphere, the eddies rotate clockwise in the Northern Hemisphere and counterclockwise in the Southern Hemi-

Figure 109   *The ocean currents.*

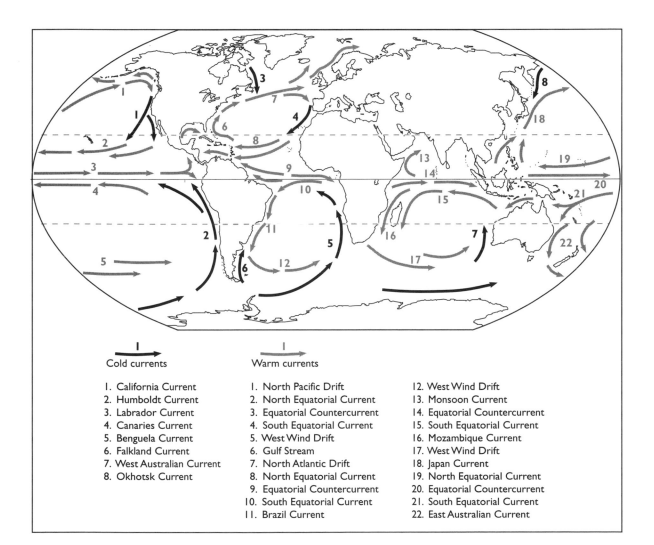

Cold currents

Warm currents

1. California Current
2. Humboldt Current
3. Labrador Current
4. Canaries Current
5. Benguela Current
6. Falkland Current
7. West Australian Current
8. Okhotsk Current

1. North Pacific Drift
2. North Equatorial Current
3. Equatorial Countercurrent
4. South Equatorial Current
5. West Wind Drift
6. Gulf Stream
7. North Atlantic Drift
8. North Equatorial Current
9. Equatorial Countercurrent
10. South Equatorial Current
11. Brazil Current

12. West Wind Drift
13. Monsoon Current
14. Equatorial Countercurrent
15. South Equatorial Current
16. Mozambique Current
17. West Wind Drift
18. Japan Current
19. North Equatorial Current
20. Equatorial Countercurrent
21. South Equatorial Current
22. East Australian Current

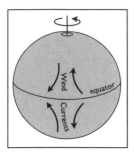

Figure 110 *The Coriolis effect causes equatorward currents to flow to the west and poleward currents to flow to the east due to the planet's rotation.*

sphere. Sea life caught in the eddies is often transported to hostile environments. Species can survive only as long as the eddies with their more favorable waters continue to operate, perhaps upward of several months.

The ocean is filled nearly to the top with icy water only a few degrees above freezing that was chilled while at the surface of the polar seas. The sinking of cold, dense water near the poles generates strong, deep currents (Fig. 111) that flow steadily toward the equator. Associated with these currents are eddies on the western side of ocean basins that are often more than 100 times stronger than the main current.

In the polar regions, the surface water is denser than in other parts of the world because of its low temperature and high salt content. The increased saltiness results from the evaporation of poleward flowing water and the exclusion of salt from ice as it freezes. As seawater increases in density, it sinks to the bottom, then spreads out upon hitting the ocean floor, and heads toward the equator. The Coriolis effect deflects the currents westward because of Earth's eastward rotation. The distribution of landmasses and the topography of the ocean floor, including ridges and canyons, also control the path taken by the circulating water.

The waters surrounding Antarctica are the coldest in the world, often below freezing. A thermal barrier produced by the circum-Antarctic current impedes the inflow of warm ocean currents. The sea ice covering the ocean around Antarctica remains for at least 10 months of the year, during which time the continent is in nearly total darkness for four months. The sea ice is punctured in various places by coastal and ocean polynyas, which are large

Figure 111 *The ocean conveyor belt, transporting warm and cold water over Earth.*

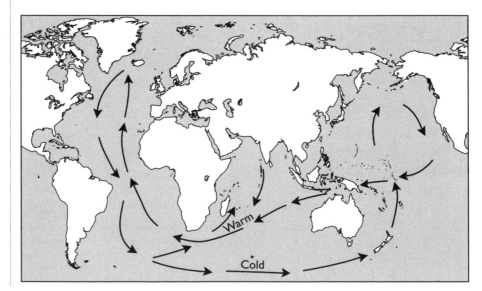

open-water areas kept from freezing by upwelling warm water currents. The coastal polynyas are essentially sea ice factories because they expose portions of open ocean that later freeze, thereby continuing the ice-making process.

The Antarctic plays a larger role in global ocean circulation than does the Arctic. Deep, cold currents flowing from Antarctica toward the equator trend to the left and press against the western side of the Atlantic, Pacific, and Indian Ocean basins. As they sweep against the continents, the currents pick up speed similar to the way a stream flows faster in a narrowing channel. The renewal of deep waters by sinking surface waters near Antarctica appears to have slowed the currents' flow dramatically from what it was more than a century ago, with dire implications for the climate. More drastic variations in ocean circulation occurred during a sharp cold snap 11,000 years ago and again during the 500-year Little Ice Age, which ended around 1880.

Swiftly flowing currents along some parts of the ocean bottom are still much of a mystery. One deep current after traveling 7,500 miles from its source in the Antarctic turns and sweeps along the edge of the abyssal plain south of Nova Scotia. The Atlantic Bottom Water, the largest mass of deep water in the world, sinks from the surface near Antarctica and flows northward along the seafloor into the western North Atlantic.

Before mixing with North Atlantic water and dispersing, some of this flow curves to the west due to Earth's eastward rotation. It hugs the lower edge of the continental rise at the border of the abyssal plain. The current, along with the lower reaches of intense eddies pinched off the Gulf Stream, could account for the muddy waters kicked up from the nearly 2,000-foot-deep abyss south of Nova Scotia that extends as far south as the Bahamas. Deep eddies induced by the Gulf Stream might be superimposed on this flow to produce undersea storms of sediment-laden water.

The Indian Ocean is unique because it is not in contact with the north-polar region and has only one source of cold bottom water from the Antarctic. By contrast, the Atlantic and Pacific connect with both the Antarctic and the Arctic Oceans. The narrow, shallow Bering Strait that separates Alaska from Asia blocks the flow of deep, cold water from the Arctic Ocean into the Pacific. Seawater freezes more readily in the Arctic regions because the near-surface water is not sufficiently enriched in salt and therefore not dense enough to sink. Consequently, the Arctic Ocean is largely a sea of ice.

Seawater in the Atlantic is saltier than in the Pacific because of the greater contribution of river-borne salt. The Atlantic has two major sources of highly saline water. One is the Gulf of Mexico, which is carried northward by the Gulf Stream. The other is the deep flow from the Mediterranean Sea. The climate of the Mediterranean is so warm that evaporation concentrates salt in seawater. Mediterranean water spilling westward through the Strait of Gibraltar sinks to a depth of nearly 4,000 feet in the Atlantic. These sources raise the salt content of the surface waters of the North Atlantic to levels much higher than in the North Pacific.

As the surface water of the North Atlantic moves northward, it enters the Norwegian Sea. There it cools below the freezing point of freshwater. However, due to its high salt content, it does not freeze. The cold, dense water sinks. Upon reaching the bottom, it reverses direction and flows back into the Atlantic through a series of narrow, deep troughs in the submarine ridges that connect Greenland, Iceland, and Scotland. This deep-sea current, called the North Atlantic Deep Water, is a subsurface stream whose flow is 20 times greater than all the world's rivers combined.

As this large volume of deep water moves southward, it flows to the right against the continental margin of eastern North America, forming the Western Boundary Undercurrent. This current transports some 20,000 cubic miles of water annually along the East Coast of North America. These deep-ocean currents travel very slowly, completing the journey from the poles to the equator and back again in upward of 1,000 years. In contrast, surface currents complete the circuit around an ocean basin in less than a decade.

The volume of rising water in parts of the ocean matches the volume of sinking water in the polar regions. The cold waters from the polar seas rise in upwelling zones in the tropics, creating an efficient heat transport system. Offshore and onshore winds also drive upwelling and downwelling currents (Fig. 112). Upwelling currents off the coasts of continents and near the equator are

Figure 112 Upwelling and sinking ocean currents are driven by offshore and onshore winds.

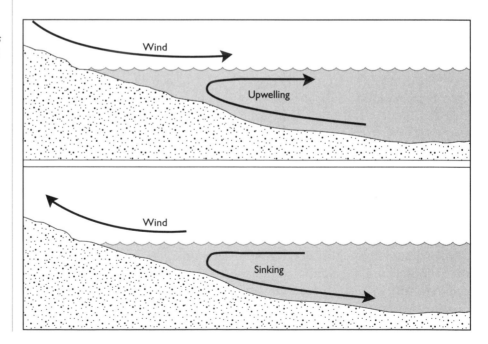

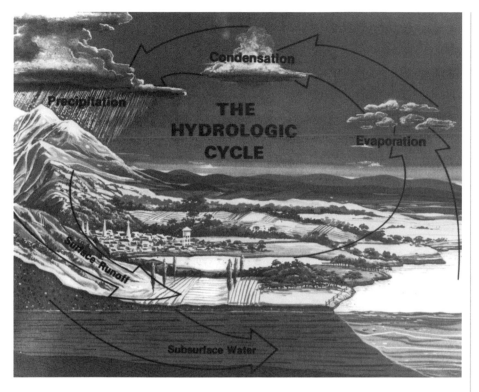

Figure 113 *The hydrologic cycle involves the flow of water from the ocean onto the land and back into the sea.*

(Photo courtesy USGS)

important sources of bottom nutrients. Modern fishers track down these areas of upwelling water, which is usually where the fish are.

Steady onshore winds also cause rip currents or riptides. Because a wave is rarely of uniform height, water piles up near the shore where incoming waves are the tallest. Water sliding down the sides of the wave creates flows that move parallel to the shore. Where the flows meet, water is diverted seaward in a narrow channel, creating a rip current that appears as a flow of choppy or dirty water. The strong currents can push out to sea unwary swimmers, who must swim parallel to the beach until out of the current and then swim back to shore.

The tropical seas are warmed by solar radiation from above and cooled by upwelling water from below. This interaction gives rise to an equator-to-pole cycle of heat transport. The interaction of the ocean and the atmosphere is also responsible for the hydrologic cycle (Fig. 113), which is the continuous flow of water from the ocean onto the land and back into the sea. These processes involve the entire ocean in a gigantic thermal engine, transporting a tremendous amount of heat around the globe.

151

EL NIÑO

Ocean currents have a dramatic effect on the weather. Changes in these systems can send abnormal weather patterns around the world. Many unusual weather events are caused by El Niño (Fig. 114), an anomalous warming of the equatorial eastern Pacific Ocean every two to seven years, with a duration of up to two years. Longer and stronger El Niños have occurred during the last two decades than in the previous 120 years. An odd double El Niño recurring consecutively manifested itself from 1991–1993 and again from 1994–1995. A powerful El Niño from 1997–1998 produced record-breaking warmth in the Pacific, causing 23,000 deaths and $33 billion in damages around the world. Seawater temperatures climbed as high as 5 degrees Celsius above normal compared with just a couple degrees for normal El Niños.

Anomalous atmospheric pressure changes in the South Pacific, called an El Niño Southern Oscillation, cause the westward-flowing trade winds to collapse. As atmospheric pressure rises on Easter Island in the eastern Pacific, it falls in Darwin, Australia, in the western Pacific. When a major El Niño occurs, the barometric pressure over the eastern Pacific falls while the pressure over the western Pacific rises. When El Niño ends, the pressure difference between these two areas reverses, creating a massive seesaw effect of atmospheric pressure.

While the El Niño is in process, it disrupts the westward-flowing trade winds. Warm water piled up in the western Pacific by the winds then flows back to the east, creating a great sloshing of water in the South Pacific

Figure 114 *Typical northern winter temperature and precipitation patterns during El Niño warming in the central Pacific. Hatched areas are dry; stippled areas are wet; and circled areas are warm.*

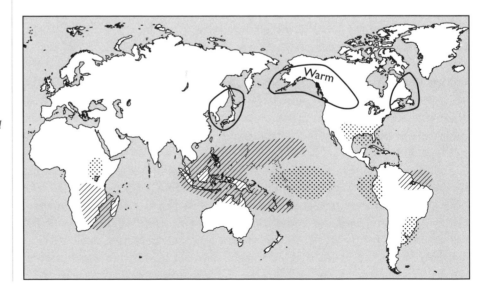

basin. The layer of warm water in the eastern Pacific becomes thicker, which suppresses the thermocline, the boundary between cold and warm water layers, and prevents the upwelling of cold water from below. This temporarily disrupts the upwelling of nutrients, which adversely affects the local marine biology.

Unusual oceanographic conditions during the 1982–1983 El Niño dramatically affected the Galápagos Islands in the Pacific, off the coast of Ecuador. The ocean current patterns around the islands are complex. They are greatly influenced by the equatorial undercurrent, a subsurface, eastward-flowing current about 600 feet thick. During a period when the sea surface temperatures were anomalously high, a major redistribution of phytoplankton around the Galápagos Islands might have contributed to the reproductive failure of seabirds and marine mammals on the islands.

A similar type of warming happens in the Indian Ocean at the same time an El Niño occurs in the Pacific. The Indian El Niño starts when warming along East Africa moves along the equator to the center of the ocean. The shifting currents are synchronized with the relocation of warmth from the western Pacific to the eastern Pacific. After El Niño has matured, the central Indian warmth moves southeast into the Timor Sea separating Australia from Indonesia. The discovery of an Indian El Niño much closer to the African continent might help explain long-distance effects such as droughts in India and southern Africa on the other side of the globe.

The opposite condition results during a La Niña, when the surface waters of the Pacific cool and are accompanied by above-average precipitation in many parts of the world. In mid-1988, water temperatures in the central Pacific plummeted to abnormally cold levels. This signaled a climate swing from an El Niño to a La Niña. The changing climate seriously affected the precipitation-evaporation balance of the world (Fig. 115). Strong monsoons hit India and Bangladesh, and heavy rains visited Australia during that year. The La Niña might also have been responsible for a severe drought in the United States during 1988 and a marked drop in global temperatures in 1989. The unusually strong El Niño of 1993 was responsible in large part for the Midwest floods, the costliest in the nation's history. The lack of either El Niño or La Niña in the winter of 2000–2001 might have been responsible for record cold conditions in many parts of the country from Arctic fronts.

Along the west coast of South America, the southeast trade winds drive the Peru current, pushing surface water offshore and allowing cold, nutrient-rich water to well up to the surface. The westward push of the trade winds continues across the eastern and central Pacific. The resulting stress on the sea surface piles up water in the western Pacific, causing the warm surface layer of the ocean to thicken in the west and thin in the east. The thermocline then falls to about 600 feet in the western Pacific and rises to about 150 feet in the

Figure 115 *The precipitation-evaporation balance of Earth. In light areas, precipitation exceeds evaporation. In dark areas, evaporation exceeds precipitation.*

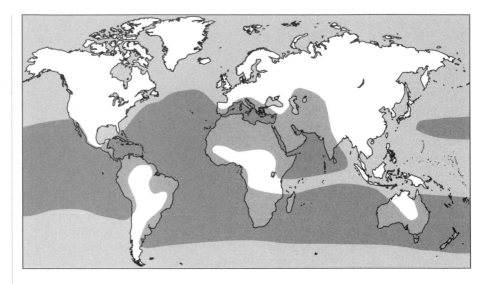

eastern Pacific. Because the thermocline is near the surface, the upwelling waters off the coast of South America are usually cold.

During an El Niño event beginning around October, the cessation of the trade winds in the western Pacific causes the thick layer of warm water to collapse. The water flows back toward the east in subsurface waves called Kelvin waves that reach the coast of South America in two to three months, creating a huge reverse flow of seawater in the South Pacific basin. The Kelvin waves generate eastward-flowing currents that transport warm water from the west. This lowers the level of the thermocline and prevents the upwelling of cool water from below.

With an increase of warm water from the west and the suppression of cold water from below, the sea surface begins to warm considerably by December or January. The stretch of warm water shifts the position of thunderstorms that pump heat and water into the atmosphere, thus rerouting atmospheric currents around the world (Fig. 116). As the El Niño continues to develop, the trade winds begin blowing from the west, intensifying the Kelvin waves and further depressing the thermocline off South America.

The Peru current, flowing northward along the west coast of South America, is not significantly weakened by the El Niño. It continues to pump water to the surface, though this time the upwelling water is warm, causing a major decline of fisheries. The westward current off equatorial South America is not only weakened by the eastward push of the Kelvin waves but is much warmer than before. This spreads the warming of the sea surface westward along the equator. The normal wind pattern reverses, causing a major disruption in global weather patterns.

Increasing surface temperatures resulting from doubling the amount of atmospheric carbon dioxide could adversely affect global precipitation patterns, especially during El Niño. The enhanced frequency of El Niño events might also be a symptom of greenhouse gas pollution and global warming. During El Niño, the oceans absorb more carbon dioxide than usual, but the land surface emits more of the gas, which tend to balance each other.

The number of weather-related natural disasters has doubled since the middle of the 20th century. In the United States, 38 severe weather events occurred from 1988 to 1999. Seven severe events occurred in 1998 alone, the most for any year on record. That was also the warmest year since global records began in 1860. Furthermore, the strongest El Niño on record occurred from 1997 to 1998, bringing Pacific seawater temperatures as high as 5 degrees Celsius above normal.

To determine whether greenhouse gases are actually warming the planet, scientists are studying the speed at which sound waves travel through the ocean. Since sound travels faster in warm water than in cold water—a phenomenon known as acoustic thermometry—long-term measurements could reveal whether global warming is a certainty. The idea is to send out low-frequency sound waves from a single station and monitor them from several listening posts scattered around the world. The signals take several hours to reach the most-distant stations. Therefore, shaving a few seconds off the

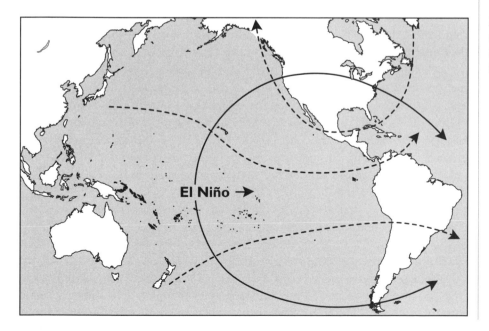

Figure 116 *Changes in air currents during El Niño. Dashed lines are normal currents.*

travel time over an extended period of 5 to 10 years could definitely indicate that the oceans are indeed warming.

ABYSSAL STORMS

The dark abyss at the bottom of the ocean was thought to be quiet and almost totally at rest, with sediments slowly raining down and accumulating at a rate of about 1 inch in 20 centuries. Recent discoveries reveal signs that infrequent undersea storms often shift and rearrange the sedimentary material that has rested for long periods on the bottom. Occasionally, the surging bottom currents scoop up the top layer of mud, erasing animal tracks and creating ripple marks in the sediments, much like those produced by wind and river currents.

On the western side of the ocean basins, undersea storms skirt the foot of the continental rise, transporting huge loads of sediment and dramatically modifying the seafloor. The storms scour the ocean bottom in some areas and deposit large volumes of silt and clay in others. The energetic currents travel at about 1 mile per hour. However, because of the considerably higher density of seawater, they sweep the ocean floor just as effectively as a gale with winds up to 45 miles per hour erodes shallow areas near shore.

The abyssal storms seem to follow certain well-traveled paths, indicated by long furrows of sediment on the ocean floor (Fig. 117). The scouring of the seabed and deposition of thick layers of fine sediment results in much more complex marine geology than that developed simply from a constant rain of sediments. The periodic transport of sediment creates layered sequences that look similar to those created by strong windstorms in shallow seas, with overlapping beds of sediment graded into different grain sizes.

Sedimentary material deposited onto the ocean floor consists of detritus, which is terrestrial sediment and decaying vegetation, along with shells and skeletons of dead microscopic organisms that flourish in the sunlit waters of the top 300 feet of the ocean. The ocean depth influences the rate of marine-life sedimentation. The farther the shells descend, the greater the chance of dissolving in the cold, high-pressure waters of the abyss before reaching the bottom. Preservation also depends on rapid burial and protection from the corrosive action of the deep-sea water.

Rivers carry detritus to the edge of the continent and out onto the continental shelf where marine currents pick up the material. When the detritus reaches the edge of the shelf, it falls to the base of the continental rise under the pull of gravity. Approximately 25 billion tons of continental material reach the mouths of rivers and streams annually. Most of this detritus is deposited near the river outlets and onto continental shelves. Only a few billion tons fall into the deep sea. In addition to the river-borne sediment, strong desert winds

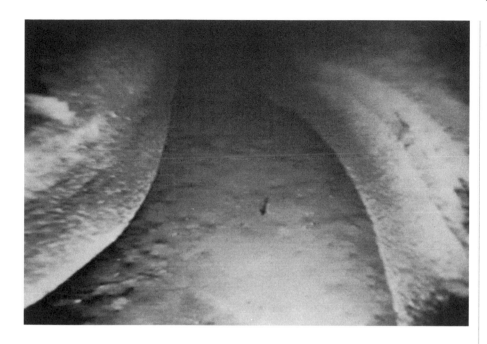

Figure 117 *A wide, flat furrow on the seabed of the Atlantic Ocean.*

(Photo by N. P. Edgar, courtesy USGS)

in subtropical regions sweep out to sea a significant amount of terrestrial material. The windblown sediment also contains significant amounts of iron, an important nutrient that supports prolific blooms of plankton. In iron-deficient parts of the ocean, "deserts" exist where "jungles" should have been even though plenty of other nutrients are available.

The biologic material in the sea contributes about 3 billion tons of sediment to the ocean floor each year. The biologic productivity, controlled in large part by the ocean currents, governs the rates of accumulation. Nutrient-rich water upwells from the ocean depths to the sunlit zone, where microorganisms ingest the nutrients. Areas of high productivity and high rates of accumulation normally occur near major oceanic fronts, such as the region around Antarctica. Other areas are along the edges of major currents, such as the Gulf Stream that circulates clockwise around the North Atlantic basin and the Kuroshio or Japan current that circles clockwise around the North Pacific basin.

The greatest volume of silt and mud and the strongest bottom currents are in the high latitudes of the western side of the North and South Atlantic. These areas have the highest potential for generating abyssal storms that form and shape the seafloor. They also have the largest drifts of sediment on Earth, covering an area more than 600 miles long, 100 miles wide, and more than 1 mile thick. Abyssal currents at depths of 2 to 3 miles play a major role in shaping the entire continental rise off North and South America. Elsewhere in the world, bottom currents shape the distribution of fine-grained material along the edges of Africa, Antarctica, Australia, New Zealand, and India.

Instruments lowered to the ocean floor measure water dynamics and their effects on sediment mobilization (Fig. 118). During abyssal storms, the velocity of bottom currents increases from about $\frac{1}{10}$ to more than 1 mile per hour. The storms in the Atlantic seem to derive their energy from surface eddies that emerge from the Gulf Stream. While the storm is in progress, the suspended sediment load increases tenfold, and the current is able to carry about 1 ton of sediment per minute for long distances. The moving clouds of suspended sediment appear as coherent patches of turbid water with a residence time of about 20 minutes. The storm itself might last from several days to a few weeks, at the end of which the current velocity slows to normal and the sediment drops out of suspension.

Not all drifts are directly attributable to abyssal storms. Material carried by deep currents has modified vast areas of the ocean as well. The storm's main effect is to stir sediment that bottom currents pick up and carry downstream for long distances. The circulation of the deep ocean does not show a strong seasonal pattern. Therefore, the onset of abyssal storms is unpredictable and likely to strike an area every 2 to 3 months.

TIDAL CURRENTS

Tides result from the pull of gravity of the Moon and Sun on the ocean. The Moon revolves around Earth in an elliptical orbit and exerts a stronger pull when on the near side of its orbit around the planet than on the far side. The difference between the gravitational attraction on both sides is about 13 percent, which elongates the center of gravity of the Earth-Moon system. The pull of gravity creates two tidal bulges on Earth. As the planet revolves, the oceans flow into the two tidal bulges, one facing toward the Moon and the other facing away from it. Between the tidal bulges, the ocean is shallower, giving it an overall egg-shaped appearance. The middle of the ocean rises only about 2.5 feet at maximum high tide. However, due to a sloshing-over effect and the configuration of the coastline, the tides on the coasts often rise several times higher.

The daily rotation of Earth causes each point on the surface to go into and out of the two tidal bulges once a day. Thus, as Earth spins into and out of each tidal bulge, the tides appear to rise and fall twice daily. The Moon also orbits Earth in the same direction it rotates, only faster. By the time a point on the surface has rotated halfway around, the tidal bulges have moved forward with the Moon, and the point must travel farther each day to catch up with the bulge. Therefore, the actual period between high tides is 12 hours, 25 minutes.

Figure 118 *Instrument to measure water dynamics and sediment mobilization on the ocean floor.*

(Photo by N. P. Edgar, courtesy USGS)

If continents did not impede the motion of the tides, all coasts would have two high tides and two low tides of nearly equal magnitudes and durations each day. These are called semidiurnal tides and occur at places such as along the Atlantic coasts of North America and Europe. However, different tidal patterns form when the tide wave is deflected and broken up by the continents. Because of this action, the tidal wave forms a complicated series of crests and troughs thousands of miles apart. In some regions, the tides are coupled with the motion of large nearby bodies of water. As a result, some areas, such as along the coast of the Gulf of Mexico, have only one tide a day called a diurnal tide, with a period of 24 hours, 50 minutes.

The Sun also raises tides with semidiurnal and diurnal periods of 12 and 24 hours. Because the Sun is much farther from Earth, its tides are only about half the magnitude of lunar tides. The overall tidal amplitude, which is the difference between the high-water level and the low-water level, depends on the relation of the solar tide to the lunar tide. It is controlled by the relative positions of the Earth, Moon, and Sun (Fig. 119).

The tidal amplitude is at maximum twice a month during the new and full moon, when the Earth, Moon, and Sun align in a nearly straight-line configuration, known as syzygy, from the Greek word *syzygos,* meaning "yoked together." This is the time of the spring tides, from the Saxon word *springan,* meaning "a rising or swelling of water," and has nothing to do with the spring season. Neap tides occur when the amplitude is at a minimum during the first and third quarters of the Moon, when the relative positions of the Earth, Moon, and Sun form at a right angle to one another and the solar and lunar tides oppose each other.

Mixed tides are a combination of semidiurnal and diurnal tides such as those that occur along the Pacific coast of North America. They display a diurnal inequality with a higher-high tide, a lower-high tide, a higher-low tide, and a lower-low tide each day. Some deep-draft ships on the West Coast must often wait until the higher of the two high tides comes in before departing. A few places, such as Tahiti, have virtually no tide because they lie on a node, a stationary point about which the standing wave of the tide oscillates.

High tides that generally exceed a dozen feet are called megatides. They arise in gulfs and embayments along the coast in many parts of the world. Megatides depend on the shape of the bays and estuaries, which channel the wavelike progression of the tide and increase its amplitude. Their height depends on the shapes of bays and estuaries, which channel the tides and increase their amplitude. Many locations with extremely high tides also experience strong tidal currents. A tidal basin near the mouth of a river can actually resonate with the incoming tide. The oscillation makes the water at one side of the basin high at the beginning of the tidal period, low in the middle, and high again at the end of the tidal period. The incoming tide sets the

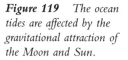

Figure 119 *The ocean tides are affected by the gravitational attraction of the Moon and Sun.*

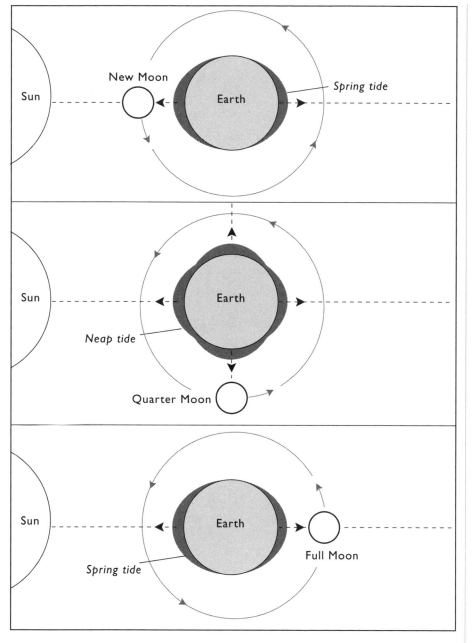

water in the basin oscillating, sloshing back and forth. The motion of the tide moving in toward the mouth of the river and the motion of the oscillation are synchronized. This reinforces the tide in the bay and makes the high tide higher and the low tide lower than they would be otherwise.

Tidal bores (Table 14) are a special feature of this type of oscillation within a tidal basin. They are solitary waves that carry tides upstream usually during a new or full moon. One of the largest tidal bores sweeps up the Amazon River. Waves up to 25 feet high and several miles wide reach 500 miles upstream. Although any body of water with high tides can generate a tidal bore, only half of all tidal bores are associated with resonance in a tidal basin. Therefore, the tides *and* their resonance with the oscillation in a tidal basin provide the energy for the tidal bore.

The seaward ends of many rivers experience tides. At the river mouth, the tides are symmetrical, with ebb and flood tide lasting about six hours each. Ebb and flood tides refer to the currents associated with the tides. Ebb currents flow out to sea, while flood currents flow into an inlet. Upstream, the tides become increasingly asymmetrical, with less time elapsing between low water and high water than between high water and low water as the tide comes in quickly but goes out gradually with the river current. A tidal bore exaggerates this asymmetry because the tide comes up the river very rapidly in a single wave.

The incoming tide arrives in a tidal basin as rapidly moving waves with long wavelengths. As the waves enter the basin, they are confined at both the sides and the bottom by the narrowing estuary. Because of this funneling action, the height of the wave increases. As the tidal bore moves upstream, it must move faster than the river current. Otherwise, it is swept downstream and out to sea.

OCEAN WAVES

Ocean waves form by large storms at sea when strong winds blow across the water's surface (Fig. 120). The wave fetch is the distance over which the wind blows on the surface of the ocean and depends on the size of the storm and the width of the body of water. For waves to reach a fully developed sea state, the fetch must be at least 200 miles for a wind of 20 knots, 500 miles for a wind of 40 knots, and 800 miles for a wind of 60 knots (a knot is 1 nautical mile per hour or 1.15 miles per hour).

The wind speed and duration determine the wave height. With a wind speed of 30 miles per hour, for example, a fully developed sea is attained in 24 hours, with wave heights up to 20 feet. The maximum sea state occurs when waves reach their maximum height, usually after three to five days of strong, steady storm winds blowing across the surface of the ocean. However, if the sustained wind blew at 60 miles per hour, a fully developed sea would have wave heights averaging more than 60 feet.

TABLE 14 MAJOR TIDAL BORES

Country	Tidal Basin	Tidal Body	Known Bore Location
Bangladesh	Ganges	Bay of Bengal	
Brazil	Amazon	Atlantic Ocean	
	Capim		Capim
	Canal Do Norte		
	Guama		
	Tocantins		
	Araguari		
Canada	Petitcodiac	Bay of Fundy	Moncton
	Salmon		Truro
China	Tsientang	East China Sea	Haining to Hangchow
England	Severn	Bristol Channel	Framilode to Gloucester
	Parrett		Bridgewater
	Wye		
	Mersey	Irish Sea	Liverpool to Warrington
	Dee		
	Trent	North Sea	Gunness to Gainsborough
France	Seine	English Channel	Gaudebec
	Orne		
	Coueson	Gulf of St. Malo	
	Vilaine	Bay of Biscay	
	Loire		
	Gironde		Îles de Margaux
	Dordogne		La Caune to Brunne
	Garonne		Bordeaux to Cadillac
India	Narmada	Arabian Sea	
	Hooghly	Bay of Bengal	Hooghly Pt. to Calcutta
Mexico	Colorado	California Gulf	Colorado Delta
Pakistan	Indus	Arabian Sea	
Scotland	Solway Firth	Irish Sea	
	Forth		
United States	Turnagain Arm	Cook Inlet	Anchorage to Portage
	Knik Arm		

Figure 120 *Open ocean waves and a mysterious weather phenomenon known as sea smoke 150 miles east of Norfolk, Virginia.*

(Photo courtesy U.S. Navy)

The wave height, measured from the top of the crest to the bottom of the trough, is generally less than 20 feet. Occasionally, storm waves of 30 to 50 feet high have been reported, but these do not occur very frequently. Exceptionally large ocean waves are rare. One such wave reported in the Pacific by a U.S. Navy tanker in 1933 was more than 100 feet high. Another large wave buckled the flight deck of the USS *Bennington* during a typhoon in the western Pacific in 1945 (Fig. 121).

The wave shape (Fig. 122) varies with the water depth. In deep water, a wave is symmetrical, with a smooth crest and trough. In shallow water, a wave is asymmetrical, with a peaked crest and a broad trough. If the water depth is more than one-half the wave length, the waves are considered deep-water waves. If the water depth is less than one-half the wave length, the waves are called shallow-water waves.

The wave length (Fig. 123) is measured from crest to crest and depends on the location and intensity of the storm at sea. The average lengths of storm waves vary from 300 to 800 feet. As waves move away from a storm area, the longer waves move ahead of the storm and form swells that travel great distances. In the open ocean, swells of 1,000-foot wave lengths are common,

Figure 121 *The buckled flight deck of the* USS Bennington *during a typhoon in the western Pacific in June 1945.*

(Photo courtesy U.S. Navy)

with a maximum of about 2,500 feet in the Atlantic and about 3,000 feet in the Pacific.

The wave period is the time a wave takes to pass a certain point and is measured from one wave crest to the next. Wave periods in the ocean vary from less than a second for small ripples to more than 24 hours. Waves with periods of less than 5 minutes are called gravity waves and include the wind-driven waves that break against the coastline, most of which have periods

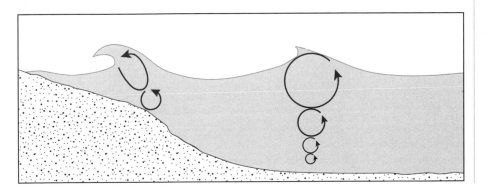

Figure 122 *The mechanics of a breaker, whose wave shape is controlled by the water depths.*

Figure 123   *Properties of waves:* (L) *wave length,* (H) *wave height,* (D) *wave depth.*

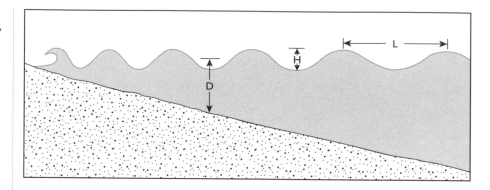

between five and 20 seconds. A seismic sea wave from an undersea earthquake or landslide usually has a period of 15 minutes or more and a wave length of up to several hundred miles.

Waves with periods between five minutes and 12 hours are called long waves and are generated by storms. Other long waves result from seasonal differences in barometric pressure over various parts of the ocean such as the Southern Oscillation discussed earlier. Waves of longer periods travel faster than shorter-period waves, with the speed proportional to the square root of the wave length. Short-period waves are relatively steep and particularly dangerous to small boats because the bow might be on a crest while the stern is in a trough, causing them to capsize or be swamped.

SEISMIC SEA WAVES

Destructive waves also result from undersea and nearshore earthquakes. They are called seismic sea waves or tsunamis, a Japanese word meaning "harbor waves," so named because of their common occurrence in this region. The waves are often referred to as tidal waves but actually have nothing to do with the tides. The vertical displacement of the ocean floor during earthquakes causes the most destructive tsunamis, whose wave energy is proportional to the intensity of the quake. The earthquake sets up ripples on the ocean similar to those formed by tossing a rock into a quiet pond.

In the open ocean, the wave crests are up to 300 miles long and usually less than 3 feet high. The waves extend downward for thousands of feet, as far as the ocean bottom. The distance between crests, or wave length, is 60 to 120 miles. This gives the tsunami a very gentle slope, which allows it to pass by ships practically unnoticed. Tsunamis travel at speeds between 300 and 600 miles per hour. Upon entering shallow coastal waters, tsunamis have been known to grow into a wall of water up to 200 feet high, although most are only a few tens of feet high.

When a tsunami touches bottom in a harbor or narrow inlet, its speed diminishes rapidly to about 100 miles per hour. The sudden breaking action causes seawater to pile up. The wave height is magnified tremendously as waves overtake one another, decreasing the distance between them in a process called shoaling. The destructive force of the wave is immense, and the damage it causes as it crashes to shore is considerable. Large buildings are crushed with ease, and ships are tossed up and carried well inland like toys (Fig. 124).

Ninety percent of all tsunamis in the world occur in the Pacific Ocean, 85 percent of which are the products of undersea earthquakes. Between 1992 and 1996, 17 tsunami attacks around the Pacific killed some 1,700 people. The

Figure 124 *Tsunamis washed many vessels into the heart of Kodiak from the March 27, 1964, Alaskan Earthquake.*

(Photo courtesy USGS)

Hawaiian Islands are in the paths of many damaging tsunamis. Since 1895, 12 such waves have struck the islands. In the most destructive tsunami, 159 people died in Hilo on April 1, 1946 by killer waves generated by a powerful earthquake in the Aleutian Islands to the north.

The March 27, 1964, Alaskan earthquake, the largest recorded to hit the North American continent, devastated Anchorage and surrounding areas. The 9.2 magnitude quake cause destruction over an area of 50,000 square miles and was felt throughout an area of half a million square miles. A 30-foot-high tsunami generated by the undersea earthquake destroyed coastal villages around the Gulf of Alaska (Fig. 125), killing 107 people. Kodiak Island was heavily damaged. Most of the fishing fleet was destroyed when the tsunami carried many vessels inland. As a striking example of the tsunami's great power, large spruce trees were snapped off with ease by a large tsunami near Shoup Bay.

The sudden change in seafloor terrain triggers tsunamis when the seabed rapidly sinks or rises during an earthquake. This either lowers or raises an enormous mound of water, stretching from the seafloor to the surface. The mound of water thrust above normal sea level quickly collapses under the pull of gravity. The vast swell can cover up to 10,000 square miles, depending on the area uplifted on the ocean floor. This alternating swell and collapse spreads out in concentric rings on the surface of the ocean.

Figure 125 Seismic sea wave damage at railroad marshaling yard, Seward district, Alaska, from the March 27, 1964, earthquake.

(Photo courtesy USGS)

Explosive eruptions associated with the birth or the death of a volcanic island also set up large tsunamis that are highly destructive. Volcanic eruptions that develop tsunamis are responsible for about one-quarter of all deaths caused by tsunamis. The powerful waves transmit the volcano's energy to areas outside the reach of the volcano itself. Large pyroclastic flows of volcanic fragments into the sea or landslides triggered by volcanic eruptions produce tsunamis as well. Coastal and submarine slides also generate large tsunamis that can overrun parts of the adjacent coast. One of the best examples was wave damage on Cenotaph Island and the south shore of Lituya Bay, Alaska, from a massive rockslide in 1958 (Fig. 126).

Large parts of Alaska's Mount St. Augustine (Fig. 127) have collapsed and fallen into the sea, generating large tsunamis. Massive landslides have ripped out the flanks of the volcano 10 or more times during the past 2,000

Figure 126 Wave damage on Cenotaph Island and the south shore of Lituya Bay, Alaska, from a massive rockslide in 1958.

(Photo by D. J. Miller, courtesy USGS)

Figure 127 *Mount St. Augustine, Cook Inlet region, Alaska.*

(Photo by C. W. Purington, courtesy USGS)

years. The last slide occurred during the October 6, 1883, eruption, when debris on the flanks of the volcano crashed into the Cook Inlet. The slide sent a 30-foot tsunami to Port Graham 54 miles away that destroyed boats and flooded houses.

In the past, earthquakes on the ocean floor went largely undetected. The only warning people had of a tsunami was a rapid withdrawal of water from the shore. Residents of coastal areas frequently stricken by tsunamis heed this warning and head for higher ground. Several minutes after the sea retreats, a tremendous surge of water extends hundreds of feet inland. Often a succession of surges occurs, each followed by a rapid retreat of water back to the sea. On coasts and islands where the seafloor rises gradually or is protected by barrier reefs, much of the tsunami's energy is spent before reaching the shore. On volcanic islands that lie in very deep water, such as the Hawaiian Islands, or where deep submarine trenches lie immediately outside harbors, an oncoming tsunami can build to prodigious heights.

Destructive tsunamis generated by large earthquakes can travel completely across the Pacific Ocean. The great 1960 Chilean earthquake of 9.5 magnitude elevated a California-sized chunk of land about 30 feet and created a 35-foot tsunami that struck Hilo, Hawaii, over 5,000 miles away, causing more than $20 million in property damages and 61 deaths. The tsunami traveled an additional 5,000 miles to Japan and inflicted considerable destruction on the coastal villages of Honshu and Okinawa, leaving 180 people dead or missing. In the Philippines, 20 people were killed. Coastal areas of New

Zealand were also damaged. For several days afterward, tidal gauges in Hilo could still detect the waves as they bounced around the Pacific basin.

Tsunami reporting stations administered by the National Weather Service are stationed in various parts of the Pacific, which is responsible for most recorded tsunamis. When an earthquake of 7.5 magnitude or greater occurs in the Pacific area, the epicenter is plotted and the magnitude is calculated. A tsunami watch is put out to all stations in the network. The military and civilian authorities concerned are notified as well. Each station in the network detects and reports the sea waves as they pass in order to monitor the tsunami's progress. The data is used to calculate when the wave is likely to reach the many populated areas at risk around the Pacific.

Unfortunately, the unpredictable nature of tsunamis produces many false warnings, resulting in areas being evacuated unnecessarily or residents completely ignoring the warnings altogether. One example occurred on May 7, 1986, when a tsunami predicted for the West Coast from the 7.7 magnitude Adak earthquake in the Aleutians, for some reason, failed to arrive. People ignored a similar tsunami warning in Hilo in 1960 at the cost of their lives. Not much can be done to prevent damage from tsunamis. However, when given the advance warning time, coastal regions can be evacuated successfully with minimal loss of life.

The most tsunami-prone area in the world is the Pacific Rim, which experiences the most earthquakes as well as the most volcanoes. Destructive tsunamis from submarine earthquakes can travel completely across the Pacific and reverberate through the ocean for days. A tsunami originating in Alaska could reach Hawaii in six hours, Japan in nine hours, and the Philippines in 14 hours. A tsunami originating off the Chilean coast could reach Hawaii in 15 hours and Japan in 22 hours. Fortunately, this gives residents in coastal areas enough time to take the necessary safety precautions that might spell the difference between loss of life and property.

After discussing ocean currents and related phenomena, the next chapter examines how these processes affect the seacoasts.

7

COASTAL GEOLOGY
THE ACTIVE COASTLINE

This chapter examines the processes that shape the seacoasts. The constant shifting of sediments on the surface and the accumulation of deposits on the ocean floor assure that the face of Earth continues to change over time. Seawater lapping against the shore during a severe storm causes coastal erosion. Steep waves accompanying storms at sea seriously erode sand dunes and sea cliffs. The continuous pounding of the surf also tears down most barriers erected against the rising sea.

America's once sandy beaches are sinking beneath the waves. Barrier islands and sandbars running along the East Coast and the coast of Texas are disappearing at alarming rates. Sea cliffs are eroding farther inland in California, often destroying expensive homes. Most defenses, such as seawalls erected to stop beach erosion, usually end in defeat as waves relentlessly batter the shoreline (Fig. 128).

SEDIMENTATION

Earth is a constantly evolving planet, with complex activities such as running water and moving waves. Rivers carry to the sea a heavy load of sediments

Figure 128 *Damage to a beach area by storms and high tides at Virginia Beach, Virginia.*

(Photo by K. Rice, courtesy USDA-Soil Conservation Service)

washed off the continents, continually building up the coastal regions. Sea-coasts vary dramatically in topography, climate, and vegetation. They are places where continental and oceanic processes converge to produce a landscape that is invariably changing rapidly. Coastal deserts are unique because they are areas where the seas meet the desert sands (Fig. 129).

Most sedimentary processes take place very slowly on the bottom of the ocean. The continents are mainly the sites of erosion, whereas the oceans are mostly the sites of sedimentation. Marine sediments consist of material washed off the continents. Most sedimentary rocks form along continental margins and in the basins of inland seas, such as those that invaded the interiors of North and South America, Europe, and Asia during the Mesozoic era. Areas with high sedimentation rates form deposits thousands of feet thick. Where they are exposed on the surface, individual sedimentary beds can be traced for hundreds of miles.

The formation of sedimentary rocks begins when erosion wears down mountain ranges and rivers carry the debris into the sea. The sediments originate from the weathering of surface rocks. The products of weathering include a wide range of materials, from very fine-grained sediments to huge boulders. Exposed rocks on the surface chemically break down into clays and carbonates and mechanically break down into silts, sands, and gravels.

Erosion by rain, wind, or glacial ice produces sediments that are brought to streams, which transport the loose sediment grains downstream to the sea. Angular sediment grains indicate a short time spent in transit. Rounded sediment grains, indicate severe abrasion from long-distant travel or from reworking by fast flowing streams or by pounding waves on the beach. Indeed, many sandstone formations were once beach deposits.

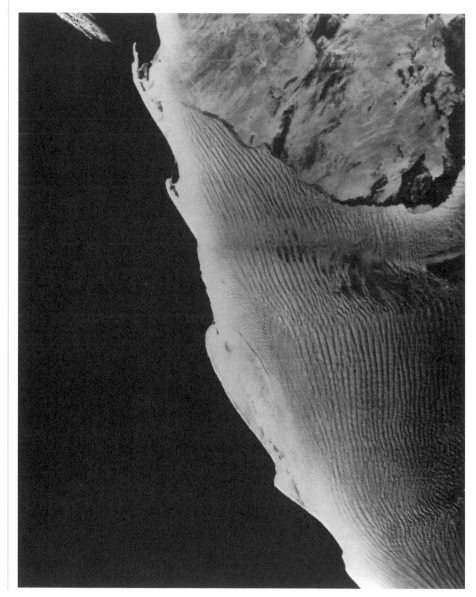

Figure 129 *Linear dunes in the northern part of the Namib Desert, Namibia, Africa.*

(Photo by E. D. McKee, courtesy USGS)

Figure 130 *Sediment deposition in the Mississippi River delta: 1930 conditions* (left), *1956 conditions* (right).

(Photo by H. P. Guy, courtesy USGS)

Annually, some 25 billion tons of sediment are carried by stream runoff into the ocean and settle onto the continental shelf. The towering landform of the Himalayas is the greatest single source of sediment. Rivers draining the region, notably the Ganges and Brahmaputra, discharge about 40 percent of the world's total amount of sediment into the Bay of Bengal, where sedimentary layers stack up to 3 miles thick.

Rivers such as the Amazon of South America and the Mississippi of North America transport enormous quantities of sediment derived from their respective continental interiors. Large-scale deforestation and severe soil erosion at its headwaters force the Amazon, the world's largest river, to carry heavier sediment loads. In addition, upland deforestation chokes off coral reefs with eroded sediments carried by rivers to the sea. The Mississippi and its tributaries drain a major section of the central United States, from the Rockies to the Appalachians. All tributaries emptying into the Mississippi have their own drainage area, forming a part of a larger basin.

Every year, the Mississippi River dumps hundreds of millions of tons of sediment into the Gulf of Mexico, widening the Mississippi Delta (Fig. 130) and slowly building up Louisiana and nearby states. The Gulf Coast states, from East Texas to the Florida panhandle, were built up with sediments eroded from the interior of the continent and hauled in by the Mississippi and other rivers. Streams, heavily laden with sediments, overflow their beds and are forced to detour as they meander toward the sea. When the streams reach the ocean, their velocity falls off sharply, and the sediment load drops out of suspension. Meanwhile, chemical solutions carried by the rivers mix thoroughly with seawater by the action of ocean waves and currents.

Upon reaching the ocean, the river-borne sediments settle out of suspension based on grain size. The course-grained sediments deposit near the turbulent shore, and the fine-grained sediments deposit in calmer waters farther out to sea. As the shoreline advances toward the sea due to the buildup of coastal sediments or a falling sea level, finer sediments are covered by progressively coarser ones. As the shoreline recedes because of the lowering of the land surface or a rising sea level, progressively finer ones cover coarser sediments.

The difference in sedimentation rates as the sea transgresses and recedes produces a recurring sequence of sands, silts, and muds. The sands comprise quartz grains about the size of beach sands, and marine sandstones exposed in the American West were deposited along the shores of ancient seas. Gravels are rare in the ocean and move mainly from the coast to the deep abyssal plains by submarine slides. In dry regions where dust storms are prevalent, the wind airlifts fine sediments out of the region. Windblown sediments landing in the ocean slowly build deposits of abyssal red clay, whose color signifies its terrestrial origin.

Cementing agents such as silica or calcium and the tremendous weight of the overlying sedimentary layers pressing down onto the lower strata lithifies the sediments into solid rock. This provides a geologic column of alternating beds of limestone, shales, siltstones, and sandstones (Fig. 131). Abrasion eventually grinds down all rocks to clay-sized particles. Because clay particles

Figure 131 A *stratigraphic cross section showing a sequence of sandstones, siltstones, and shales, overlying a basement rock composed of limestone.*

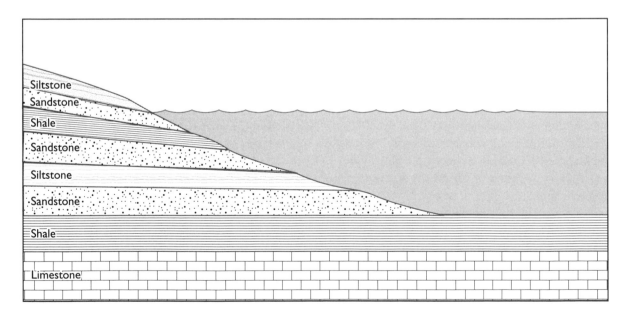

are so small and sink so slowly, they normally settle out in calm, deep waters far from shore. Compaction from the weight of the overlying strata squeezes out water between sediment grains, lithifying the clay into mudstone or shale.

The color of sedimentary beds helps identify the type of depositional environment. Generally, sediments tinted various shades of red or brown indicate a terrestrial source, whereas green- or gray-colored sediments suggest a marine origin. The size of individual particles influences the color intensity, and darker colored sediments usually indicate finer grains.

The varying thicknesses of sediment layers reflect different depositional environments at the time they were laid down. Each bedding plane marks where one type of deposit ends and another begins. Thick sandstone beds might be interspersed with thin beds of shale. This indicate periods of coarse sediment deposition punctuated by periods of fine sediment deposition as the shoreline progressed and receded.

Graded bedding occurs when particles in a sedimentary bed vary from coarse at the bottom to fine at the top. This type of bedding indicates the rapid deposition of sediments of differing sizes by a fast flowing stream emptying into the sea. The largest particles settle out first and, due to the difference in settling rates, are covered by progressively finer material. Beds also grade laterally, producing a horizontal gradation of sediments from coarse to fine.

Limestones were laid down on the bottoms of oceans or large lakes. Some formations were once ancient coral reefs. Limestones are among the most common rocks and make up about 10 percent of the land surface. They are composed of calcium carbonate mostly derived from biologic activity as evidenced by abundant fossils of marine life in limestone beds. Coquina is a limestone consisting almost entirely of fossils or their fragments. Some limestones chemically precipitate directly from seawater or deposit in freshwater lakes. Minor amounts precipitate in evaporite deposits formed in brine pools that periodically evaporate.

Chalk is a soft, porous carbonate rock. One of the largest deposits is the chalk cliffs of Dorset, England, where poor consolidation of the strata results in severe erosion during coastal storms. Dolomite resembles limestone, with the calcium in the original carbonate partially replaced by magnesium. The chemical reaction can cause a reduction in volume and create void spaces. The Dolomite Alps in northeast Italy are upraised blocks of this mineral deposited on the bottom of an ancient sea.

The sediments settle onto the continental shelf (Fig. 132), which extends up to 100 miles or more and reaches a depth of roughly 600 feet. In most places, the continental shelf is nearly flat, with an average slope of only about 10 feet per mile. Beyond the continental shelf lies the continental slope, which extends to an average depth of more than 2 miles. It has a steep angle of several degrees, comparable to the slopes of many mountain ranges.

Figure 132 *A profile of the ocean floor.*

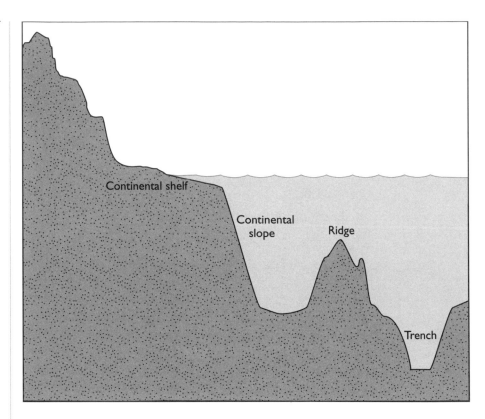

Sediments reaching the edge of the continental shelf slide down the continental slope under the pull of gravity. Often, huge masses of sediment cascade down the continental slope by gravity slides that can gouge out steep submarine canyons. They play an important role in building up the continental slope and the smooth ocean bottom below.

STORM SURGES

Storms at sea produce pressure changes and strong winds (Table 15) that pile up seawater and cause flooding when occurring at high tides. Waves generated by high winds superimposed on regular tides produce the most severe tidal floods, especially when the Moon, Sun, and Earth are in alignment. While the tide is in, high waves raise the maximum level of the prevailing high tide. Strong onshore winds blowing toward the coast push seawater onto the shore. The opposite condition occurs when strong offshore winds blow toward the ocean during low tide, lowering the sea significantly and sometimes grounding vessels in port.

Most high waves and beach erosion occur during coastal storms. Thunderstorms and squalls are the most violent storms. They are most frequent in the midlatitudes and produce gusty winds, hail, lightning, and a rapid buildup of seas. The life cycle of a single thunderstorm cell is usually less than half an hour. When the cell dies, a new one develops in its place.

Frontal storms form at the leading edge of a cold front. A squall line often precedes a cold front, with a distinctive dark gray, cylinder-shaped cloud that appears to roll across the sky from one end of the horizon to the other. Squall lines travel about 25 miles per hour, with winds reaching 60 miles per hour. However, they are generally short-lived, usually lasting less than 15 minutes. When a squall arrives, it produces waves several feet high. Since the winds do not last long, the waves die down almost as rapidly as they build up.

Hurricanes and typhoons produce the most dramatic storm surges (Fig. 133). Hurricane-force winds caused by the rotation and forward motion of the storm reach 100 miles per hour or more, pushing water out in front of the storm. The low pressure in the eye of the hurricane draws water up into a mound several feet high. As the hurricane moves across the ocean and its speed matches the speed of the waves, it often sets up a resonance with the swells it generates. This action adds to the height of the swells, which have been reported to be more than 60 feet high in some hurricanes.

TABLE 15   THE BEAUFORT WIND SCALE

Beaufort Number	Description	Miles per Hour	Indications
0	Calm	< 1	Smoke rises vertically
1	Light air	1–3	Direction of wind shown by smoke drift but not by wind vane
2	Light breeze	4–7	Wind felt on face; leaves rustle
3	Gentle breeze	8–12	Leaves and small twigs in constant motion; wind extends light flag
4	Moderate breeze	13–18	Raises dust and loose vapor; moves small branches
5	Fresh breeze	19–24	Small trees begin to sway; crested wavelets form on inland water
6	Strong breeze	25–31	Large branches in motion; telephone wires whistle
7	Near gale	32–38	Whole trees in motion; resistance when walking against the wind
8	Gale	39–46	Breaks twigs off trees; generally impedes progress
9	Strong gale	47–54	Breaks large limbs off trees; slight structural damage occurs
10	Storm	55–63	Uproots trees; considerable structural damage occurs
11	Violent storm	64–75	Widespread damage
12–17	Hurricane	> 75	Devastation occurs; storm surge damages coastal areas

Figure 133 *Overwash damage from storm surge at Cape Hatteras, North Carolina.*

(Photo by R. Dolan, courtesy USGS)

When a hurricane approaches the coast, the water piled up by the wind, the mounding of water by the low pressure, the generation of swells, and the possible resonance of swell waves can make a most deadly combination when superimposed onto the regular cycle of incoming tides. The result is massive flooding, devastation of property, and the loss of life.

Torrential downpours and tidal floods from hurricanes and typhoons cause more damage and take more lives than other forms of flooding. By their very nature, the tropical storms drop huge amounts of rainfall over large areas often within a day or so. The deluge causes widespread flooding in natural drainage areas, where streams cannot cope with the excess water formed by the onrush of heavy rains.

Tidal floods are overflows on coastal areas bordering the ocean, an estuary, or a large lake. Coastal lands, including bars, spits, and deltas, offer the same protection from the sea that floodplains do from rivers. Coastal flooding is primarily a result of high tides, waves from high winds, storm surges, tsunamis, or any combination of these. Tidal floods also occur when waves generated by hurricane winds combine with flood runoff due to heavy rains that accompany the storms.

The flooding can extend over large distances along a coastline. The duration is usually short and depends on the elevation of the tide, which usually rises and falls twice daily. If the tide is in, other forces that produce high waves can raise the maximum level of the prevailing high tide. The most severe tidal floods result when waves produced by high winds combine with the regular tides, causing a tremendous amount of damage as well as severe beach erosion that continues to move the coastline inland.

COASTAL EROSION

Coastal landslides occur when a sea cliff is undercut by wave action and falls into the ocean (Fig. 134). Sea cliff retreat is caused by marine and nonmarine agents, including wave attack, wind-driven salt spray, and mineral solution. The nonmarine agents responsible for cliff erosion include chemical and mechanical processes, surface drainage water, and rainfall. Mechanical erosion processes include cycles of freezing and thawing of water in crevices, which forces existing fractures to split even farther apart. Weathering agents break down rocks or cause the outer layers to peel or spall off. Animal trails that

Figure 134 Highway 1 at the Devils Slide, San Mateo County, California.

(Photo by R. D. Brown, courtesy USGS)

weaken soft rocks also affect sea cliff erosion, as do burrows that intersect cracks in the soil.

Surface water runoff and wind-driven rain further erode the sea cliff. Excessive rainfall along the coast can also lubricate sediments, causing huge blocks of rock to slide into the ocean. Water running over the cliff edge and wind-driven rain produce the fluting often exposed on cliff faces. Groundwater seeping from a cliff can create indentations on the cliff face, which undermine and weaken the overlying strata. The addition of water also increases pore pressure within sediments, reducing the shear strength (internal resistance to stress) that holds the rock together. If bedding planes, fractures, or jointing dip seaward, water moving along these areas of weakness can produce rockslides. This process has excavated large valleys on the windward parts of the Hawaiian Islands, where powerful springs emerge from porous lava flows.

Glacial ice also erodes coastal regions. During the ice ages, glaciers gouged deep fjords (Fig. 135). These are long, narrow, steep-sided inlets in glaciated mountainous coasts. Fjords are found in the coastal mountains of Norway, Greenland, Alaska, British Columbia, Patagonia in southern South America, and Antarctica. As a tidewater glacier on the coast eroded its valley floor below sea level, it cut a steep-walled, troughlike arm of the ocean. When sea levels returned to normal at the end of the last ice age, the ocean invaded deeply excavated glacial troughs in the coastline. The side walls along the fjord are characterized by hanging valleys and tall waterfalls.

The main type of marine erosion is direct wave attack at the base of the cliff. This quarries out weak beds and undercuts the cliff until the overlying unsupported material collapses onto the beach. Waves also work along joint or fault planes to loosen blocks of rock or soil. In addition, the wind carries salt spray from breaking waves into the air and drives it against the sea cliff. Porous sedimentary rocks absorb the salty water, which evaporates and forms salt crystals that weaken rocks. The surface of the cliff slowly flakes off and falls to the beach below. The material falls to the base of the cliff and piles up, forming a talus cone—a steep-sided pile of rock fragments.

Solution erosion attacks limestone cliffs, where chemical processes dissolve soluble minerals from the rocks. The seawater dissolves lime in the rocks, forming deep notches in the sea cliffs (Fig. 136). Chemical erosion also removes cementing agents in rocks, causing sediment grains to separate. Limestone erosion is prevalent on coral islands in the South Pacific and on the limestone coasts of the Mediterranean and Adriatic Seas.

The erosion of sea cliffs and dunes that mark the coastline causes the shoreline to retreat a considerable distance. One of the worst cases in the United

Figure 136 A tidal terrace at low tide, Puerto Rico.

(Photo by C. A. Kaye, courtesy USGS)

183

Figure 137 *A system of groins to trap sand moving laterally along the beach at Lake Michigan.*

(Photo by P.W. Koch, courtesy USDA-Soil Conservation Service)

States occurred between 1888 and 1958, when the coastline on Cape Cod, Massachusetts, from Nauset Spit to Highland Light retreated at an average rate of more than 3 feet per year. In England, the soft cliffs of the Suffolk Coast on the North Sea erode at an average rate of 10 to 15 feet per year. At the town of Lowestoft, a single storm eroded a 40-foot-high cliff of unconsolidated rock 40 feet inland. Where the cliff stood only about 6 feet high, it eroded 90 feet.

Beach erosion is difficult to predict and almost impossible to stop. It depends on the strength of beach dunes or sea cliffs, the intensity and frequency of coastal storms, and the exposure of the coast. Most attempts to prevent beach erosion are defeated because the waves constantly batter and erode defenses to keep out the sea (Fig. 137). Jetties and seawalls erected to halt the tides increase erosion dramatically. In their attempts to stabilize the seashores, developers are destroying the very beaches upon which they intend to build.

The rate of retreat varies with the shape of the shoreline and the prevailing wind and tides. More than half of the 72-mile-long south shore of Long Island, New York, is considered a high-risk zone for development, with the sea reclaiming some locations at a rate of 6 feet per year. The barrier island that runs from Cape Henry, Virginia to Cape Hatteras, North Carolina has narrowed from both the seaward and landward sides (Fig. 138). The rest of the North Carolina Coast is moving back at 3 to 6 feet annually, and most of East Texas is vanishing even faster. In California, homes are falling because of the undercutting of sea cliffs, causing considerable property damage (Fig. 139).

Figure 138 *Serious losses of property near Cape Hatteras, Dare County, North Carolina, caused by shoreline regression and storm surges.*

(Photo by R. Dolan, courtesy USGS)

Figure 139 *The erosion of these bluffs at Point Montara, California, will eventually deliver buildings, roads, and other structures to the sea.*

(Photo by R. D. Brown, courtesy USGS)

About 80 percent of America's once-sandy beaches are sinking beneath the waves. Most of the problems stem from the methods engineers use to stabilize the beaches. Jetties cut off the natural supply of sand to beaches, and seawalls increase erosion by bouncing waves back without absorbing much of their energy. The rebounding waves carry sand out to sea, undermining the beach and destroying the shorefront property the seawall was designed to protect.

In an effort to protect houses on eroding bluffs overlooking the sea, coastline residences often erect expensive seawalls. Yet these structures actually hasten the erosion of sand from the beaches in front of the wall. In effect, the seawalls are saving the bluffs at the detriment of the beaches. Barriers erected at the bottom of sea cliffs might deter wave erosion but have no effect on sea spray and other erosional processes. Although beaches in front of the seawalls often lose sand during certain seasons, waves return sand at other times.

Natural processes will not replenish the disappearing sand along beaches on the East Coast until the next ice age. Most of the sand along the coast and continental shelf originates in the north from sources such as the Hudson River. For sand to move as far south as the Carolina coast, it must progress in steps possibly taking millions of years.

As the sand moves along a coast, ocean currents push it into large bays or estuaries. The embayment continues to fill with sand until sea levels drop and the accumulated sediments flush down onto the continental shelf. The sand travels only as far as the next bay in a single glacial cycle, however. Therefore, most beaches would not receive a major restocking of sand until the next ice age.

WAVE IMPACTS

Large storms at sea generate most ocean waves when strong winds blow across the surface of the water. Waves breaking along the coast dissipate energy and are responsible for generating alongshore currents, which in turn transport sand along the beach. The currents are detected using a Doppler radar called Ocean Surface Current Radar (OSCR), which determines the speed of water flow. The radar system uses two transmitters stationed several miles apart along a beach. They send out radio signals over the water and receive echoes reflected back by ocean waves to produce a map of currents in the region. OSCR can help scientists understand beach erosion, the health of fisheries, and the movement of pollution through the water.

Waves also cause coastal erosion, a serious problem in areas where the shoreline is steadily receding. Most beach erosion from high waves occurs during coastal storms. On large lakes or bays, sudden barometric pressure changes cause the water to slosh back and forth, producing waves called seiches. They are common on Lake Michigan and, on occasions, can be quite destructive. Hurricanes produce the most dramatic storm surges, which are responsible for destroying entire beaches. As a wave approaches the shore, it touches bottom and slows. The shoaling of the wave distorts its shape, causing it to break upon the beach. The breaking wave dissipates its energy along the coast and causes beach erosion.

Wave reflection bounces wave energy off steep beaches or seawalls and is responsible for the formation of sandbars. When waves approach the shore at an angle to the beach, the wave crests bend by refraction. When waves pass the end of a point of land or the tip of a breakwater, a circular wave pattern generates behind the breakwater. The refracted waves intersect other incoming waves, increasing the wave height.

Wave steepness is the ratio of wavelength to wave height and is one of the most important aspects of waves. Storm waves with high steepness have short wavelengths and high wave heights, and they produce choppy seas. Steep waves accompanying storms at sea cause erosion of sea cliffs and sand dunes along the coast. Swells with low steepness generally result in the shoreward transport of sediment. Therefore, much of the sediment carried offshore by storm waves returns by swells during the interval between storms.

As waves leave the storm area, they develop into swells that travel great distances, sometimes halfway around the world before dying out or intercepting a coastline. As the waves spread outward from the storm area, the longer-period waves move out in front while the shorter-period waves trail behind. As swells move across the ocean toward distant shores, the low, long-period waves arrive first, followed by higher swells with shorter periods.

Waves expanding outward from a storm center form rings similar to those produced by tossing a rock into a quiet pond. As the rings enlarge, the wave spreads out along a greater length, expanding the circumference of the circle. This increases the wave height as it moves away from the storm area. When swells arrive at the coast, they generate a uniform succession of waves, each with about the same period and height, which change when the slower swells begin to arrive.

The wave motion changes as waves travel from deep water toward the shore. The waves transport energy but not the water itself. As the wave crest approaches, an object floating on the surface first rises and moves forward with the crest, drops into the trough, and then moves backward. Thus, a floating object describes a circular path with the diameter equal to the wave height and returns to its original position after the wave passes.

When swells reach a coast, they form various types of breakers (Fig. 140), depending on the wave steepness and bottom slope conditions near the beach. If the slope is relatively flat, less then 3 degrees, the wave forms a spilling breaker, the most common type. This is an oversteepened wave that starts to

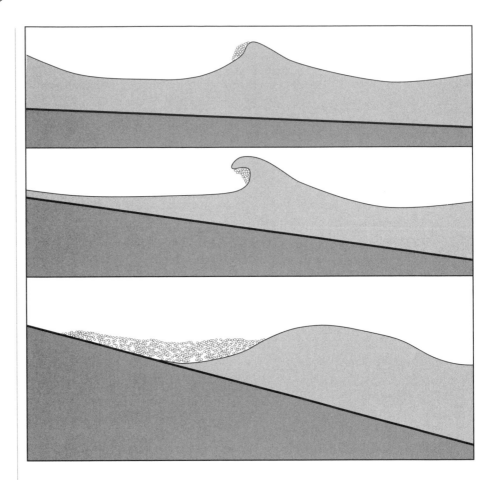

break at the crest and continues breaking as the wave travels toward the beach, providing good waves for surfing.

A plunging breaker forms when the bottom slope is between 3 and 11 degrees and the crest curls over, forming a tube of water. As the wave breaks, the tube moves toward the shore bottom and stirs up sediments. Plunging waves are the most dramatic breakers and do the most beach damage because the energy concentrates at the point where the wave breaks.

A collapsing breaker forms when the bottom slope is between 11 and 15 degrees. The breaker is confined to the lower half of the wave. However, as the wave moves toward the coast, most of it reflects off the beach.

A surging wave develops on a steep bottom where the slope is greater than 15 degrees. The wave does not break but surges up the beach face and reflects off the coast, generating standing waves near the shore. Standing waves are important for the development of offshore structures such as bars, sand spits, beach cusps, and riptides.

COASTAL SUBSIDENCE

Coastal subsidence often occurs during large earthquakes that cause one block of crust to drop below another. Vegetated lowlands along the coast are elevated by the influx of sediments to avoid inundation by the sea. When an earthquake strikes, these lowlands sink far enough to be submerged regularly and become barren tidal mudflats (Fig. 141). Between earthquakes, sediments fill the tidal

Figure 141   *A tidal flat at the head of Bolinas Lagoon, showing secondary cracks, Marin County, California.*

(Photo by G. K. Gilbert, courtesy USGS)

flats and raise them to the level where vegetation can grow once again. Therefore, repeated earthquakes produce alternating layers of lowland soil and tidal flat mud.

Earthquake-induced subsidence in the United States has occurred mainly in California, Alaska, and Hawaii. The subsidence results from vertical displacements along faults that can affect broad areas. During the 1964 Good Friday, Alaska, earthquake, more than 70,000 square miles of land tilted downward more than 3 feet, causing extensive flooding in coastal areas of southern Alaska. Flow failures usually develop in loose saturated sands and silts. They originate on land and on the seafloor near coastal areas. The Alaskan earthquake produced submarine flow failures that destroyed seaport facilities at Valdez, Whittier, and Seward. The flow failures also generated large tsunamis that overran coastal areas and caused additional casualties.

Some of the most spectacular examples of nonseismic subsidence in the United States are along coasts (Fig. 142). The Houston-Galveston area in Texas has experienced local subsidence of as much as 7.5 feet and subsidence of 1 foot or more over an area of 2,500 miles, mostly from the withdrawal of groundwater. In Galveston Bay, the ground subsided 3 feet or more over an area of several square miles following oil extraction from the underlying strata. Subsidence in some coastal towns has increased susceptibility to flooding during severe coastal storms.

The pumping of large quantities of oil at Long Beach, California, caused the ground to subside, forming a huge bowl up to 25 feet deep over an area of about 20 square miles. In some parts of the oil field, land subsided at a rate of 2 feet per year. In the downtown area, the subsidence was upward of 6 feet, causing severe damage to the city's infrastructure. The injection of seawater under high pressure into the underground reservoir halted most of the subsidence, with the added benefit of increasing the production of the oil wells.

Some of the most dramatic examples of earthquake-caused subsidence are along seacoasts (Fig. 143). Coastal cities also subside due to a combination of rising sea levels and withdrawal of groundwater, causing the aquifer to compact. Subsidence in some coastal areas has increased susceptibility to flooding during earthquakes or severe coastal storms. Coastal regions of Japan are particularly susceptible to subsidence. Parts of Niigata, Japan, sank below

Figure 143 Subsidence of the coast at Halape from the November 29, 1975, Kalapana earthquake, Hawaii County, Hawaii.

(Photo by R. I. Trilling, courtesy USGS)

191

Figure 144 *The Nile River Valley, viewed from the space shuttle, serves some 50 million people in a 7,500-square-mile area.*

(Photo courtesy NASA)

sea level during the extraction of water-saturated natural gas, requiring the construction of dikes to keep out the sea. During the June 16, 1964, earthquake, the dikes were breached with seawater when the city subsided 1 foot or more, causing serious flooding in the area of subsidence. A tsunami generated by the earthquake also damaged the harbor area.

The overdrawing of groundwater has caused the land to sink around building foundations in the northeastern section of Tokyo, Japan. The subsidence progressed at a rate of about 6 inches per year over an area of about 40 square miles, one-third of which sank below sea level. This prompted the construction of dikes to keep out the sea from certain sections of the city during a typhoon or an earthquake. A threat of catastrophe hangs over Tokyo from inundation by floodwaters during earthquakes and typhoons that have always plagued the region. Had the January 17, 1995, Kobe earthquake of

7.2 magnitude struck Tokyo instead, more than half the city would have sunk beneath the waves.

The Nile Delta of Egypt (Fig. 144) is heavily irrigated and supports 50 million people in a 7,500 square mile area. Port Said on the northeast coast of the delta sits at the northern entrance to the Suez Canal. The region overlies a large depression filled with 160 feet of mud, indicating that part of the delta is slowly dropping into the sea. Over the last 8,500 years, this portion of the fan-shaped delta has been lowering by less than one-quarter inch per year. However, more recently, the yearly combined subsidence and sea level rise have greatly exceeded this amount, which could place major portions of the city underwater. Moreover, as the land subsides, seawater infiltrates into the groundwater system, rendering it useless.

Many coastal cities subside because of a combination of rising sea levels and withdrawal of groundwater, which causes compaction of the aquifer beneath the city. Generally, the amount of subsidence is on the order of 1 foot for every 20 to 30 feet of lowered water table. Underground fluids fill inter-granular spaces and support sediment grains. The removal of large volumes of fluid, such as water or petroleum, results in a loss of grain support, a reduction of intergranular void spaces, and the compaction of clays. This action causes the land surface to subside wherever widespread subsurface compaction occurs (Fig. 145).

Over the last 50 years, the cumulative subsidence of Venice, Italy, has been about 5 inches. The Adriatic Sea has risen about 3.5 inches over the last century, resulting in a relative sea level rise of more than 8 inches. The severe subsidence causes Venice to flood during high tides, heavy spring runoffs, and storm surges.

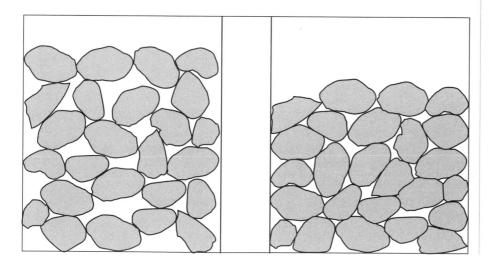

Figure 145 *The subsidence of sediments* (right) *by the withdrawal of fluids.*

MARINE TRANSGRESSION

Sea levels have risen and fallen many times throughout geologic history. More than 30 rises and falls of global sea levels occurred between 6 and 2 million years ago. At its highest point between 5 and 3 million years ago, the global sea level rose about 140 feet higher than today. Between 3 and 2 million years ago, the sea level dropped at least 65 feet lower than at present due to growing glaciers at the poles. During the ice ages, sea levels dropped as much as 400 feet at the peak of glaciation. Global sea levels steadied about 6,000 years ago after rising rapidly for thousands of years following the melting of the great glaciers that sprawled across the land during the last ice age.

Civilizations have had to endure changing sea levels for centuries (Table 16). If the ocean continues to rise, the Dutch who reclaimed their land from the sea would find a large portion of their country lying underwater. Many islands would drown or become mere skeletons of their former selves with only their mountainous backbones showing above the water. Half the scattered islands of the Republic of Maldives southwest of India would be lost. Much of Bangladesh would also drown, a particularly distressing situation

TABLE 16   MAJOR CHANGES IN SEA LEVEL

Date	Sea Level	Historical Event
2200 B.C.	Low	
1600 B.C.	High	Coastal forest in Britain inundated by the sea.
1400 B.C.	Low	
1200 B.C.	High	Egyptian ruler Ramses II builds first Suez canal.
500 B.C.	Low	Many Greek and Phoenician ports built around this time are now under water.
200 B.C.	Normal	
A.D. 100	High	Port constructed well inland of present-day Haifa, Israel.
A.D. 200	Normal	
A.D. 400	High	
A.D. 600	Low	Port of Ravenna, Italy becomes landlocked. Venice is built and is presently being inundated by the Adriatic Sea.
A.D. 800	High	
A.D. 1200	Low	Europeans exploit low-lying salt marshes.
A.D. 1400	High	Extensive flooding in low countries along the North Sea. The Dutch begin building dikes.

since the heavily populated region seriously floods during typhoons. Because they are located on seacoasts or along inland waterways, the seas would inundate most of the major cities of the world, with only the tallest skyscrapers poking above the waterline. Coastal cities would have to rebuild farther inland or construct protective seawalls to hold back the waters.

The global sea level appears to have risen upward of 9 inches over the last century due mostly to the melting of the polar ice caps. The present rate of sea level rise is several times faster than half a century ago, amounting to about 1 inch every five years. The melting of the polar ice caps due to a sustained warmer climate increases the risk of coastal flooding around the world during high tides and storms. The additional freshwater in the North Atlantic could also affect the flow of the Gulf Stream, causing Europe to freeze while the rest of the world continues to warm. The calving of large numbers of icebergs from glaciers entering the ocean could substantially raise sea levels, thereby drowning coastal regions. Consequently, beaches and barrier islands inevitably disappear as shorelines move inland (Fig. 146).

Figure 146 *Old stumps and roots exposed by shore erosion at Dewey Beach, Delaware, indicate that this area was once the tree zone.*

(Photo by J. Bister, courtesy USDA-Soil Conservation Service)

The present rate of melting is comparable to the melting rate of the continental glaciers at the end of the last ice age. The rapid deglaciation between 16,000 and 6,000 years ago, when torrents of meltwater entered the ocean, raised the sea level on a yearly basis only a few times greater than it is rising today. Higher sea levels are also caused in part by sinking coastal lands due to the increased weight of seawater pressing down onto the continental shelf. In addition, sea level measurements are affected by the rising and sinking of the land surface due to plate tectonics and the rebounding of the continents after glacial melting at the end of the last ice age.

As global temperatures increase, coastal regions where half the people of the world live would feel the adverse effects of rising sea levels due to melting ice caps and thermal expansion of the ocean. In areas such as Louisiana, the sea level has risen upward of 3 feet per century, increasing the risk of beach wave erosion (Fig. 147). The thermal expansion of the ocean has also raised the sea level about 2 inches. Surface waters off the California coast have warmed nearly 1 degree Celsius over the past half century, causing the water to expand and raise the sea about 1.5 inches.

If all the polar ice melted, the additional seawater would move the shoreline up to 70 miles inland in most places. The rising waters would inundate

Figure 147 *Beach wave erosion at Grand Isle, Louisiana.*

(Photo courtesy Army Corps of Engineers)

low-lying river deltas that feed much of the world's population. The inunda-tion would radically alter the shapes of the continents. The receding shores would result in the loss of large tracks of coastal land along with shallow bar-rier islands. All of Florida along with south Georgia and the eastern Carolinas would vanish. The Gulf Coastal plain of Mississippi, Louisiana, East Texas, and major parts of Alabama and Arkansas would virtually disappear. Much of the isthmus separating North and South America would sink out of sight.

At the present rate of melting, the sea could rise 1 foot or more by the middle of the century. For every foot of sea level rise, 100 to 1,000 feet of shoreline would be inundated, depending on the slope of the coast. Just a 3-foot rise could flood about 7,000 square miles of coastal land in the United States, including most of the Mississippi Delta, possibly reaching the outskirts of New Orleans.

The current sea level rise is upward of 10 times faster than a century ago, amounting to about one-quarter inch per year. Most of the increase appears to result from melting ice caps, particularly in West Antarctica and Greenland. Greenland holds about 6 percent of the world's freshwater in its ice sheet. An apparent warming climate is melting more than 50 billion tons of water a year from the Greenland ice sheet, amounting to more than 11 cubic miles of ice annually. In addition, higher global temperatures could influence Arctic storms, increasing the snowfall in Greenland 4 percent with every 1 degree Celsius rise in temperature.

About 7 percent of the yearly rise in global sea level results from the melting of the Greenland ice sheet and the calving of icebergs from glaciers entering the sea (Fig. 148). The Greenland ice sheet is undergoing significant thinning of the southern and southeastern margins, in places as much as 7 feet a year. Furthermore, Greenland glaciers are moving more rapidly to the sea. This is possibly caused by meltwater at the base of the glaciers that helps lubri-cate the downhill slide of the ice streams. In an average year, some 500 ice-bergs spawn from western Greenland and drift down the Labrador coast, where they become shipping hazards. In 1912, the oceanliner *Titanic* was sunk by such an iceberg.

Most of the ice flowing into the sea from the Antarctic ice sheet dis-charges from a small number of fast-moving ice streams and outlet glaciers. The grounding line is the point where the glacier reaches the ocean and the ice lifts off the bedrocks and floats as an iceberg. More icebergs are calving off glaciers entering the sea. They appear to be getting larger as well, threatening the stability of the ice sheets. The number of extremely large icebergs has also increased dramatically. Much of this instability is blamed on global warming.

One of the largest known icebergs separated from the Ross Ice Shelf in late 1987 and measured about 100 miles long, 25 miles wide, and 750 feet thick, about twice the size of Rhode Island. In August 1989, it collided with

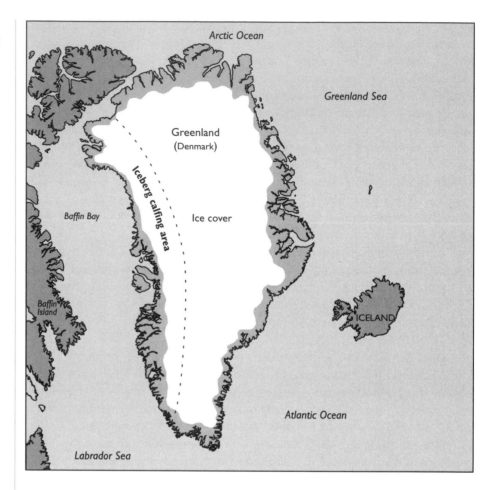

Antarctica and broke in two. Another extremely large iceberg measuring 48 miles by 23 miles broke off the floating Larson Ice Shelf in early March 1995 and headed into the Pacific Ocean. The northern portion of the Larson Ice Shelf, located on the east coast of the Antarctic Peninsula, has been rapidly disintegrating, which accounts for such gargantuan icebergs.

Perhaps during the biggest icebreaking event in a century, an iceberg about 180 miles long and 25 miles wide (or roughly the size of Connecticut) split off from the Ross Ice Shelf in early spring 2000. The breaking off of the iceberg is most likely part of the normal process of ice shelf growth and not necessarily a consequence of global warming. These giant icebergs could pose a serious threat if they drift into the Ross Sea and block shipping lanes to McMurdo Station 200 miles away.

Alpine glaciers also contain substantial quantities of ice. Many mountaintop glaciers are rapidly melting, possibly due to a warmer climate. Some areas such as the European Alps might have lost more than half their cover of ice. Moreover, the rate of loss appears to be accelerating. Tropical glaciers such as those in the high mountains of Indonesia have receded at a rate of 150 feet per year over the last two decades. At the present rate of temperature rise and rate of retreat, the glaciers are likely to disappear completely.

Sea ice covers most of the Arctic Ocean to a thickness of 12 feet or more and forms a frozen band of thinner ice around Antarctica (Fig. 149) during the winter season in each hemisphere. These polar regions are most sensitive to global warming and experience greater atmospheric changes than other parts of the world. About half of Antarctica is bordered by ice shelves. The two largest, the Ross and Filchner-Ronne, are nearly the size of Texas. The 2,600-foot-thick Filchner-Ronne Ice Shelf might actually thicken with global warming, which would enhance the ice-making process. Many other ice shelves could become unstable and float freely in a warmer climate. Since the 1950s, several smaller ice shelves have disintegrated, and today some larger shelves are starting to retreat.

A period known as stage II, a warm interlude between ice ages around 400,000 years ago, was a 30,000-year-period of global warming that eclipsed that of today. During this time, the melting of the ice caps caused the sea level to rise about 60 feet higher than at present. Most of the high seas were caused

Figure 149 *U.S. Coast Guard icebreaker* Polar Star *near Palmer Peninsula, Antarctica.*

(Photo by E. Moreth, courtesy U.S. Navy)

by the melting of the West Antarctic ice shelves, leaving open ocean in their place. The rest came from the melting of the stable East Antarctic ice cap and the Greenland ice sheet.

The present interglacial could become equally as warm if not warmer than stage II if average global temperatures continue to rise at their present rate. The warmer climate could induce an instability in the West Antarctic ice sheet, causing it to surge into the sea. This rapid flow of ice into the ocean could raise sea levels up to 20 feet or more, inundate the continents several miles inland, and flood valuable property. In the United States alone, a full quarter of the population would find itself underwater, mostly along the East and Gulf Coasts. If all the ice on Antarctica, which holds 90 percent of the world's total, were to melt, enough water would be dumped into the ocean to raise global sea levels nearly 200 feet.

Other factors contributing to rising sea levels are the extraction of groundwater, redirection of rivers for agriculture, drainage of wetlands, deforestation, and other activities that divert water to the oceans, all of which account for about one-third of the global sea level rise. When water stored in aquifers, lakes, and forests is released at a faster rate than it is replaced, the water eventually ends up in the oceans. Forests store water in both their living tissues and the moist soil shaded by plant cover. Also, one of the products of combustion when forests are burned is water. When forested areas are destroyed, the water within eventually winds up in the ocean, thus raising the sea level.

Most countries would feel the adverse effects of rising sea levels as increasing sea temperatures cause the ice caps to melt. If the melting continues at its present rate, the sea could rise 6 feet by the middle of this century. Large tracks of coastal land would disappear along with shallow barrier islands and coral reefs. Low-lying fertile deltas that support millions of people would drown. Delicate estuaries, where many species of marine life hatch their young, would be reclaimed by the ocean. Vulnerable coastal cities would have to relocate farther inland or build costly seawalls to protect against the rising waters.

After discussing coastal processes, the next chapter deals with the natural resources provided by the sea, including energy, minerals, and nutrition.

8

SEA RICHES
RESOURCES OF THE OCEAN

This chapter examines the bounty of the sea—its energy and mineral potential. The world is fortunate to have such an abundance of natural resources (Table 17), which have dramatically advanced civilization. Much of this wealth comes from the sea, which holds the key to unheard-of riches. Hidden in the world's oceans are untouched reserves of petroleum and minerals along with huge fisheries that provide half the dietary protein requirements for the human race.

The capacity of the oceans to generate energy surpasses all fossil fuels combined. The harnessing of this vast energy source could meet the demand for centuries to come. New frontiers for future exploration include the continental shelves and the ocean depths. Improved exploration techniques will ensure, with proper management, a continued supply of ocean resources well into the future.

LAW OF THE SEA

The United States initiated the expansion of national claims to the ocean and its resources with the Truman Proclamations on the Continental Shelf and the

TABLE 17   NATURAL RESOURCE LEVELS

DEPLETION RATE IN YEARS AT PRESENT CONSUMPTION

Commodity	Reserves*	Total Resources
Aluminum	250	800
Coal	200	3,000
Platinum	225	400
Cobalt	100	400
Molybdenum	65	250
Nickel	65	160
Copper	40	270
Petroleum	35	80

* Reserves are recoverable resources with today's mining technology.

Extended Fisheries Zone of 1945. Other nations followed this expansion of national boundaries and began carving up the world's oceans in a manner similar to the colonial division of Africa a century earlier. On December 6, 1982, 119 countries signed the United Nations Convention on the Law of the Sea. The declaration was a kind of constitution for the sea. It put 40 percent of the ocean and its bottom next to the coasts of continents and islands under the management of the states in possession of those regions. The other 60 percent of the ocean surface and the water below were reserved for the traditional freedom of the seas.

The remaining wealth of the ocean floor, or about 40 percent of Earth's surface, was deeded to the "Common Heritage of Mankind." The convention placed that heritage under the management of an International Seabed Authority, with the capacity to generate income, the power of taxation, and an eminent domain like authority over ocean-exploiting technology. The convention also provided a comprehensive global framework for protecting the marine environment, a new regime for marine scientific research, and a comprehensive legal system for settling disputes. It ensured freedom of navigation and free passage through straits used for international maritime activities, a right that cannot be suspended under any circumstances. In essence, the Law of the Sea provided a new order more responsive to the real needs of the world.

Coastal states were accorded a 12-mile limit of territorial sea and a 12-mile contiguous zone. Beyond these limits, they were granted a 200-mile economic zone (Fig. 150) that included fishing rights and rights over all resources. In cases where the continental shelf extended beyond the 200-mile limit, the economic zone with respect to resources on the seabed was expanded to 350 miles. The economic zone concept has also been described as the greatest territorial grab in history, giving coastal states unfair advantage over landlocked countries, thus increasing inequality among nations.

In March 1983, the United States added more than 3 million square miles to its jurisdiction by declaring the waters 200 miles offshore as the nation's Exclusive Economic Zone (EEZ), an area which is considerably larger than the country itself. In 1984, the British oceanographic ship *Farnella* began a six-year comprehensive mapping project of the ocean floor in the United States' EEZ for future resources of petroleum and minerals. The maps revealed features possibly overlooked by smaller-scale studies. Along the West Coast were dozens of newly discovered seamounts and earthquake faults. On the western side of the Gulf of Mexico were oil-trapping salt domes, submarine slides, and undersea channels. In addition, large sand dune fields similar to those found in the deep Pacific lay in the Gulf under 10,000 feet of water. The American research vessel *Samuel P. Lee* (Fig. 151) went on a similar mission in the Bering Sea to explore for oil and gas.

While diving along a midocean spreading center called the Gorda Ridge about 125 miles off the coast of Oregon, the U.S. Navy's deep submersible *Sea Cliff* discovered in September 1988 a lush community of exotic animals in a field of hot springs. Similar hot spring oases have been found on other spreading centers, where molten rock from the mantle rises

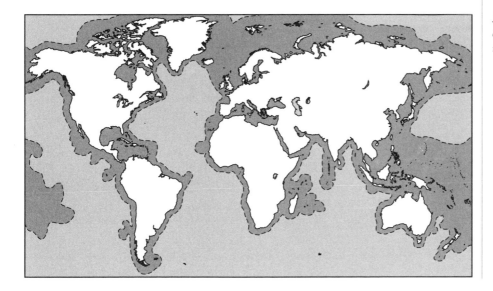

Figure 150 *The world's economic zones of marine resources.*

Figure 151 *The research vessel* Samuel P. Lee *carried out geophysical surveys in the Pacific Ocean and Alaskan waters.*

(Photo courtesy USGS)

to create new ocean crust as two adjoining crustal plates pull apart. However, this was the first hydrothermal vent system existing within the United States' EEZ. Moreover, the site might be a source for such strategic minerals as manganese and cobalt, used for strengthening steel. The hydrothermal water of up to 400 degrees Celsius often carries dissolved minerals that form deposits on the ocean floor when the hot water mixes with the near-freezing bottom water.

The discovery of a significant resource anywhere in the world's ocean could invite a claim from the nearest coastal or island state even if it lies beyond the limits of national jurisdiction. Such a dispute has occurred over a splattering of semisubmerged coral reefs in the South China Sea for their oil potential. Disputes over the ownership of midocean ore deposits have diminished the interests of western industrial nations. The future of undersea mining and refining of manganese nodules and other metallic ores is left in the hands of many Asian countries, including Japan, China, South Korea, and India, which need these resources to reduce their dependence on foreign raw materials.

The expansion of national jurisdictions into the oceans also constrains the freedom of the seas for scientific research such as core drilling on the ocean floor (Fig. 152). Under present law, other nations must apply for consent from a coastal state to conduct research in waters that were once open to all. Opposition to such a scientific project by a coastal country that controls the waters in question might undermine the cooperative atmosphere among nations that the Law of the Sea was supposed to foster.

Figure 152 *The seafloor drillship* Paul Langevin III *was used to obtain rock cores of the Juan de Fuca ridge.*

(Photo courtesy USGS)

OIL AND GAS

Of all the mineral wealth lying beneath the waves, only oil and natural gas fields in shallow coastal waters have been profitable under present economic conditions. More than 1 trillion barrels of oil have thus far been discovered, of which fully one-third or more has already been depleted. The world consumes about 70 million barrels of oil daily, with the United States using nearly one-third of the total. An average American consumes more than 40 barrels of oil a year compared with the average European or Japanese who uses between 10 and 30 barrels annually. In contrast, an average person in a developing country uses the equivalent of only one or two barrels of oil yearly.

Petroleum provides nearly half the world's energy, with about 20 percent of the oil and about 5 percent of the natural gas production offshore. In the future, perhaps half the world's petroleum will be extracted from the seabed. Unfortunately, much offshore oil leaks into the oceans, amounting up to 2 million tons each year. Such pollution could become an enormous environmental problem as production increases to keep up with demand.

Over the last two decades, offshore drilling for oil and natural gas in shallow coastal waters has become extremely profitable. Interest in offshore oil began in the mid-1960s. A considerable increase in drilling occurred a decade later following the 1973 Arab oil embargo, when American motorists stood in long lines at gas stations. New important finds such as Prudhoe Bay on Alaska's North Slope (Fig. 153) and on the North Sea off Great Britain came out of intensive exploration for new reserves of offshore oil.

In the early 1980s, the Department of the Interior estimated that 27 billion barrels of oil and 163 trillion cubic feet of natural gas remain to be discovered in offshore deposits large enough to be commercially exploited around the United States. Estimates of undiscovered oil resources are by their very nature uncertain and are based largely on geologic data. After four years of intense exploration, however, the department cut in half its estimates of oil reserves in offshore fields. The new figures reflected the fact that oil companies came up with nearly 100 dry wells after drilling in highly promising areas of the Atlantic and off the coast of Alaska.

The desire for energy independence encouraged oil companies to explore for petroleum in the deep oceans. There they encountered many difficulties, including storms at sea and the loss of personnel and equipment. Such difficulties and problems could not justify the few discoveries that were made. Futuristic plans foresee building drilling equipment and workrooms on the seafloor where they are not affected by storms. This would make some deep-sea oil and gas fields available for the first time.

To test whether people can live successfully undersea for extended periods, the National Oceanic and Atmospheric Administration (NOAA) oper-

Figure 153 *An oil tanker approaches the Valdez terminal of the trans-Alaskan pipeline, bringing North Slope petroleum to the lower 48 states.*

(Photo courtesy U.S. Maritime Administration)

ated a subsea laboratory called Aquarius. The underwater laboratory, situated in the Florida Keys, was outfitted with automated life-support systems, advanced computers, and communication equipment. High-resolution color video images, voice transmissions, and data from the lab traveled by cable to a buoy above and was transmitted to shore using microwave signals. An acoustic tracking system monitored the locations and air supplies of divers, who were not required to decompress between dives. Underwater scientists who have been studying Florida's coral reefs since 1993 could stay in Aquarius for as long as 10 days at a time.

The creation of reservoirs of oil and natural gas requires a special set of geologic circumstances, including a sedimentary source for the oil, a porous rock to serve as a reservoir, and a confining structure to trap the oil. From several tens of millions to a few hundred million years are needed to produce petroleum, depending mainly on the temperature and pressure conditions within the sedimentary basin. The source material is organic carbon trapped in fine-grained, carbon-rich sediments. Porous and permeable sedimentary rock such as sandstones and limestones form the reservoir. Geologic structures created by folding or faulting of sedimentary beds trap or pool the oil. Petroleum often associates with thick beds of salt. Because salt is lighter than the overlying sediments, it rises toward the surface, creating salt domes that help trap oil and natural gas.

The organic material that forms petroleum originates from microscopic organisms living primarily in the surface waters of the ocean and that concentrate in fine particulate matter on the ocean floor. The transformation of organic material into oil requires a high rate of accumulation or a low oxygen content in the bottom water to prevent oxidation of organic material before burial under layers of sediment. Oxidation causes decay, which destroys organic matter. Therefore, areas with high rates of accumulation of sediments rich in organic material are the most favorable sites for the formation of oil-bearing rock. Deep burial in a sedimentary basin heats the organic material under high temperatures and pressures, which chemically alters it. Essentially, the organic material is "cracked" into hydrocarbons by the heat generated in Earth's interior. If the hydrocarbons are overcooked, natural gas results.

The hydrocarbon volatiles locked up in the sediments along with seawater migrate upward through permeable rock layers and accumulate in traps formed by sedimentary structures that provide a barrier to further migration. In the absence of such a cap rock, the volatiles continue rising to the surface and escape into the ocean from natural seeps, amounting to about 1.5 million barrels of oil yearly. This amount is minuscule, however, compared with some 25 million barrels of oil a year accidentally spilled into the ocean (Fig. 154).

Some oil and gas wells might actually help clean up natural pollutants leaking from the seafloor. Records of these seeps go back to the early Spanish explorers who sailed the channel off the coast of Santa Barbara, California, and noticed oil slicks on the water. As the oil ages, it transforms into a tarlike substance that washes up on shore. Pumping oil and gas offshore has decreased the amount of petroleum leaking out of the seafloor from natural seeps by about half, most dramatically near production platforms. Removal of oil and gas over the years has decreased pressure in the subsea hydrocarbon formation, thereby reducing the amount of petroleum oozing up to the seafloor. However, when fluids and gas are injected into the rock to drive up pressure to increase production, the flow of natural seeps also increases.

The geology of the ocean floor determines whether the proper conditions exist for trapping oil and gas and greatly aids oil companies in their exploration activities. Petroleum exploration begins with a search for sedimentary structures conducive to the formation of oil traps. Seismic surveys delineate these structures by using air gun explosions that generate waves similar to sound waves, which are received by hydrophones towed behind a ship (Fig. 155). The seismic waves reflect and refract off various sedimentary layers, providing a sort of geologic CAT scan of the ocean crust.

After choosing a suitable site, the oil company brings in a drilling platform (Fig. 156). This stands on the ocean floor in shallow water or floats freely while anchored to the bottom in deep water. While drilling through the bottom sediments, workers line the well with steel casing to prevent cave-ins and to act as a conduit for the oil. A blowout preventer placed on top of the casing prevents the oil from gushing out under tremendous pressure once the drill bit penetrates the cap rock. If the oil well is successful, additional wells are drilled to develop the field fully.

Figure 154 *The December 19, 1976,* Argo Merchant *oil spill off Nantucket, Massachusetts.*

(Photo courtesy NOAA)

Figure 155 *A seismic survey of the ocean's crust.*

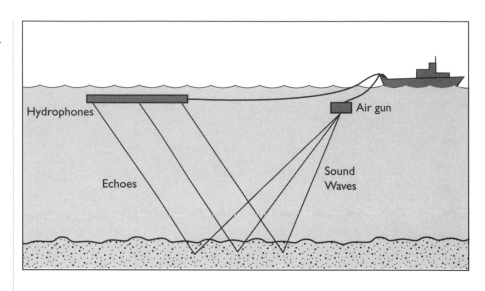

Figure 156 *A drilling platform in the Grand Isle area off of Louisiana.*

(Photo by E. F. Patterson, courtesy USGS)

Reservoirs of hot gas-charged seawater called geopressured deposits lying beneath the Gulf Coast off Texas and Louisiana are a hybrid of natural gas and geothermal energy. The gas deposits formed millions of years ago when seawater permeated porous beds of sandstone between impermeable clay layers. The seawater captured heat building up from below and dissolved methane from decaying organic matter. As more sediments piled on top of this formation, the hot gas-charged seawater became highly pressurized. Wells drilled into this formation tapped both geothermal energy and natural gas, providing an energy potential equal to about one-third of all coal deposits in the United States.

Another potential source of energy is a snowlike natural gas deposit called methane hydrate on the deep ocean floor. Methane hydrate is a solid mass formed when high pressures and low temperatures squeeze water molecules into a crystalline cage around a methane molecule. Vast deposits of methane hydrate are thought to be buried in the seabed around the continents and represent the largest untapped source of fossil fuel left on Earth. Methane hydrates hidden beneath the waters around the United States alone hold enough potential natural gas to supply all the nation's energy needs for perhaps hundreds of years.

Tapping into this enormous energy storehouse, however, is costly and potentially dangerous. If the methane hydrates become unstable, they could erupt like a volcano. Several craters on the ocean floor are identified as having been caused by gas blowouts. Giant plumes of methane have been observed rising from the seabed. Methane escaping from the hydrate layer also nourishes microbes that, in turn, sustain cold-vent creatures such as tubeworms. Additionally, methane, a potent greenhouse gas, escaping into the atmosphere could escalate global warming.

MINERAL DEPOSITS

Ores are naturally occurring materials from which valuable minerals are extracted. Miners have barely scratched the surface in their quest for ore deposits. Immense mineral resources lie at great depths, awaiting the mining technology to bring them to the surface for use in industry. Many of these deposits had their origins on the bottom of the sea (Fig. 157). Improved techniques in geophysics, geochemistry, and minerals exploration has helped keep resource supplies up with rising demand. As improved exploration techniques become available, future supplies of minerals will be found in yet unexplored regions. Precision radar altimetry from satellites and other remote sensing techniques can map the ocean bottom, where a large potential for the world's future supply of minerals and energy exists.

Figure 157 *The location of ore deposits originally formed by seafloor hot springs.*

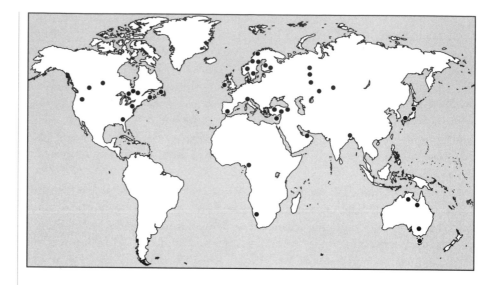

Mineral ore deposits form very slowly, taking millions of years to create an ore significantly rich to be suitable for mining. Certain minerals precipitate over a wide range of temperatures and pressures. They commonly occur together with one or two minerals predominating in sufficiently high concentrations to make their mining profitable. Extensive mountain building activity, volcanism, and granitic intrusions provide vein deposits of metallic ores.

Hydrothermal (hot-water emplaced) deposits are a major source of industrial minerals. The discovery of hydrothermal ores has stimulated intense study of their genesis for more than a century. Toward the turn of the 20th century, geologists found that hot springs at Sulfur Bank, California, and Steamboat Springs, Nevada (Fig. 158), deposited the same metal-sulfide compounds that are found in ore veins. Therefore, if the hot springs were depositing ore minerals at the surface, hot water must be filling fractures in the rock with ore as it moves toward the surface. The American mining geologist Waldemar Lindgren discovered rocks with the texture and mineralogy of typical ore veins by excavating the ground a few hundred yards from Steamboat Springs. He proved that many ore veins formed by circulating hot waters called hydrothermal fluids. The mineral fillings precipitated directly from hot waters percolating along underground fractures.

Hydrothermal ores originate when a gigantic subterranean still is supplied with heat and volatiles from a magma chamber. As the magma cools, silicate minerals such as quartz crystallize first, leaving behind a concentration of other elements in a residual melt. Further cooling of the magma causes the rocks to shrink and crack. This allows the residual magmatic fluids to escape toward the surface and invade the surrounding rocks to form veins.

The rocks surrounding a magma chamber might be another source of minerals found in hydrothermal veins, with the volcanic rocks acting only as a heat source that pumps water into a giant circulating system. Cold, heavier water moves down and into the volcanic rocks, carrying trace amounts of valuable elements leached from the surrounding rocks. When heated by the magma body, the water rises into the fractured rocks above, where it cools, loses pressure, and precipitates its mineral content into veins.

Figure 158 *Steam fumaroles at Steamboat Springs, Nevada.*

(Photo by W. D. Johnston, courtesy USGS)

Two metals on opposite extremes of the hydrothermal spectrum are mercury and tungsten. All belts of productive deposits of mercury are associated with volcanic systems. Mercury is the only metal that is liquid at room temperature. It forms a gas at low temperatures and pressures. Therefore, much of Earth's mercury is lost at the surface from volcanic steam vents and hot springs. Tungsten, by comparison, is one of the hardest metals, which makes it valuable for hardening steel. It precipitates at very high temperatures and pressures, often at the contact between a chilling magma body and the rocks it invades.

Hydrothermal ores deposited by hot water are also associated with volcanically active zones on the ocean floor. These zones include midocean ridges that create new oceanic crust and island arcs on the margins of subduction zones that destroy old oceanic crust. Hydrothermal deposits exist on young seafloors along active spreading centers of the major oceans as well as regions that are rifting apart and forming new oceans such as the Afar Rift, the Gulf of Aden, and the Red Sea (Fig. 159). In addition, deep-sea drilling has uncovered identical deposits in older ocean floors far from modern spreading centers. This suggests that the process responsible for the creation of metal deposits has operated throughout the history of the major oceans.

Rich ores, including copper, zinc, gold, and silver, lie hidden among the midocean rifts. The hydrothermal deposits form by the precipitation of minerals in hot-water solutions rich in silica and metals discharged from hydrothermal springs. Silica and other minerals build prodigious chimneys, from which turbulent black clouds of fluid (black smokers) billow out. Metal-rich particles precipitated from the effluent fill depressions on the seafloor and eventually form an ore body.

The minerals that contribute to hydrothermal systems originate from the mantle at depths of 20 to 30 miles below the seafloor. Magma upwelling from the mantle penetrates the oceanic crust and provides new crustal material at spreading centers. Seawater seeping into fractures in the basaltic rock on the ocean floor penetrates below the crust near the magma chamber. There it circulates within the zone of young, highly fractured rock and heats to a temperature of several hundred degrees Celsius.

The hot water is kept from boiling by the pressure of several hundred atmospheres. The water dissolves silica and minerals from the basalt, which are carried in solution to the surface by convection and discharged through fissures in the seafloor (Fig. 160). In addition, metal-rich fluids derived directly from the magma and volatile elements from the mantle also travel along with the hydrothermal waters to the surface. When the hot metal-rich solution emerges from a vent into cold, oxygen-rich seawater, metals such as iron and manganese are oxidized and deposit along with silica. Some deposits on the Mid-Atlantic Ridge contain as much as 35 percent manganese, an important metal used in steel alloys.

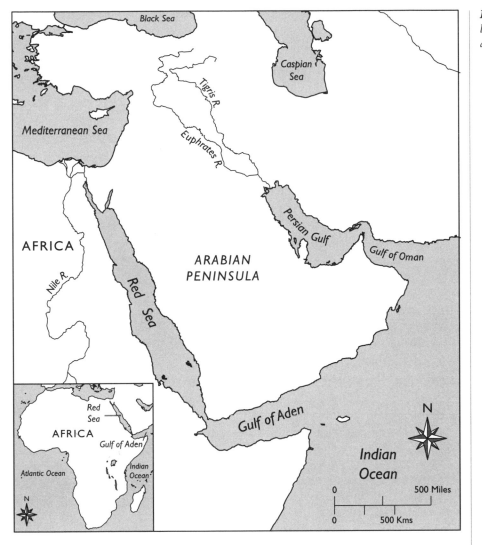

Figure 159 *The location of the Red Sea and Gulf of Aden.*

The hydrothermal deposits are generally poor in copper, nickel, cobalt, lead, and zinc because these elements remain in solution longer than iron and manganese. Under oxygen-free conditions, such as those in stagnant pools of brine, copper and zinc tend to concentrate along with iron and manganese. These deposits occur in the Red Sea, where the concentrations of copper and zinc reach ore grades sufficiently high to make mining economical.

Another type of ore deposit exists in ophiolites, which are fragments of ancient oceanic crust uplifted and exposed on land by continental collisions. The grounded oceanic crust consists of an upper layer of marine sediments, a layer of pillow lava (basalts erupted undersea), and a layer of dark, dense ultra-

Figure 160 The
operation of hydrothermal
vents on the seafloor.

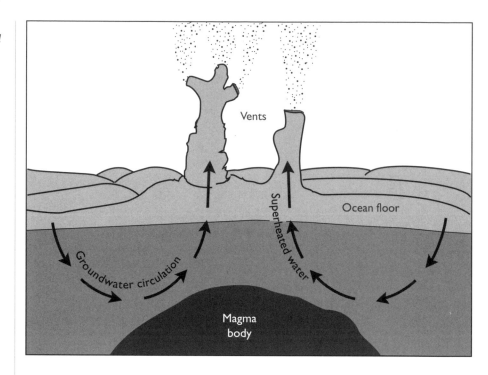

mafic (iron-magnesium-rich) rocks possibly derived from the upper mantle. The metal ore deposits exist at the base of the sedimentary layer just above the area where it contacts the basalt.

Ophiolite ore deposits are scattered throughout many parts of the world (Fig. 161). They include the 100-million-year-old ophiolite complexes

Figure 161 The
worldwide distribution of
ophiolites, which are slices
of oceanic crust shoved up
onto land by plate
tectonics.

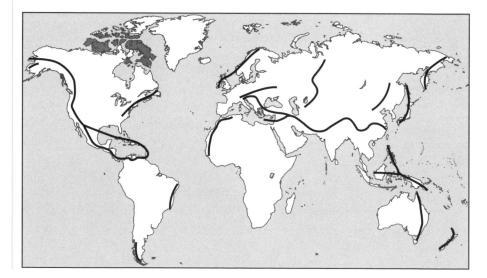

exposed on the Apennines of northern Italy, the northern margins of the Himalayas in southern Tibet, the Ural Mountains in Russia, the eastern Mediterranean (including Cyprus), the Afar Desert of northeastern Africa, the Andes of South America, the islands of the western Pacific such as the Philippines, uppermost Newfoundland, and Point Sol along the Big Sur coast of central California.

Another type of mineral ore emplacement is called massive sulfide deposits. They originated on the ocean floor at midocean spreading centers and occurred as disseminated inclusions or veins in ophiolite complexes that were exposed on dry land during continental collisions. One of the most noted deposits is in the 100-million-year-old Apennine ophiolites, which were first mined by the ancient Romans. Massive sulfide deposits are mined extensively in other parts of the world for their rich ores of copper, lead, zinc, chromium, nickel, and platinum.

Massive sulfides are metal ore deposits formed at midocean spreading centers. The sulfide metals deposited by hydrothermal systems form large mounds on the ocean floor (Fig. 162 and Fig. 163). The deposits contain sulfides of iron, copper, lead, and zinc and occur in most ophiolite complexes. That are mined extensively throughout the world for their rich ores. The circulating seawater below the ocean floor acquires sulfate ions and becomes strongly acidic. This reaction promotes the combination of sulfur with certain metals leached from the basalt and extracted from the hydrothermal solution to form insoluble sulfide minerals.

Figure 162 *A weathered sulfide mound on the Juan de Fuca ridge.*

(Photo courtesy USGS)

Figure 163 *Formation of a massive sulfide deposit by hydrothermal fluids.*

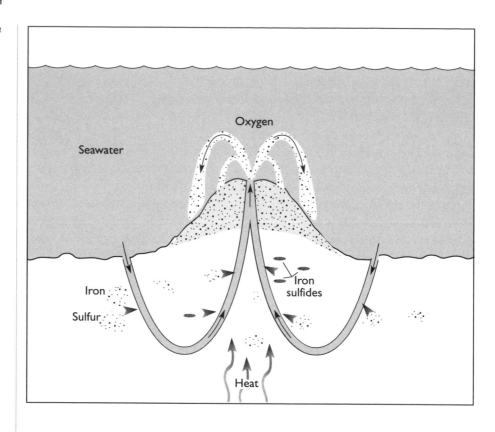

The massive sulfide deposits also occur as disseminated inclusions or veins in the rock below the seafloor in ophiolites (Fig. 164). Another deposit forms only when a ridge axis is near a landmass, which is a source of large amounts of erosional debris. The massive sulfide ore body lies in the midst of a sediment layer, usually shale derived from fine-grained clay. Some of the world's most important deposits of copper, lead, zinc, chromium, nickel, and platinum that are critical to modern industry originally formed several miles below the seafloor and upthrusted onto dry land during continental collisions.

Ore deposits are also associated with hot brines resulting from the opening of a new ocean basin by a slow spreading center, such as the one bisecting the Red Sea. Hot, metal-rich brines fill basins along the spreading zone. The cold, dense seawater percolating down through volcanic rocks becomes unusually salty because it passes through thick beds of halite (sodium chloride) buried in the crust. These salt beds formed under dry climatic conditions when evaporation exceeded the inflow of seawater in a nearly enclosed basin.

When the salinity reached the saturation point, salt crystals precipitated out of solution and settled onto the ocean floor, accumulating in thick beds. The high salinity of hot circulating solutions through these salt beds enhanced their ability to transport dissolved metals by forming complexes with the chlorine in the salt. When they discharged from the floors of the basins, the heated solutions collected as hot brines. Metals precipitated from the hot brines and settled in basins, where they formed layered deposits of metalliferous sediments up to 6 miles thick in places.

Evaporite deposits are produced in arid regions near the shore. There pools of brine, which are constantly replenished with seawater, evaporate in the hot sun, leaving salts behind. The deposits generally form between 30 degrees north and south of the equator. However, extensive salt deposits are not being formed at present, which suggests a cooler global climate. Ancient evaporite deposits existing as far north as the Arctic regions indicate that either these areas were at one time closer to the equator or the global climate was considerably warmer in the geologic past. Evaporite accumulation peaked about 230 million years ago when the supercontinent Pangaea was beginning to rift apart. Few evaporite deposits date beyond 800 million years ago, however, probably because most of the salt formed before then has been recycled back into the ocean.

The salts precipitate out of solution in stages. The first mineral to precipitate is calcite, closely followed by dolomite, although only minor amounts of limestone and dolostone are produced in this manner. After about two-thirds of the water is evaporated, gypsum precipitates. When nine-tenths of the water is removed, halite, or common salt, forms. Thick deposits of halite are also produced by the direct precipitation of seawater in deep basins that have

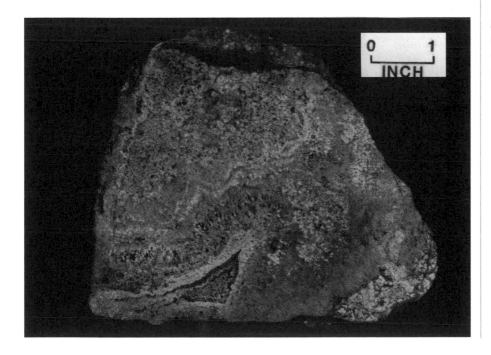

Figure 164 *A metal-rich massive sulfide vein deposit in ophiolite.*

(Photo courtesy USGS)

been cut off from the general circulation of the ocean such as the Mediterranean and the Red Seas.

Thick beds of gypsum, composed of hydrous calcium sulfate deposited in the continental interiors, constitute one of the most common sedimentary rocks. They are produced in evaporite deposits that formed when a pinched-off portion of the ocean or an inland sea evaporated. Oklahoma, as with many parts of the interior of North America that were invaded by a Mesozoic sea, is well-known for its gypsum beds. The mineral is mined extensively for the manufacture of plaster and drywall board.

Sulfur is one of the most important nonmetallic minerals. It occurs in abundance in sedimentary and evaporite deposits, with volcanoes contributing only a small proportion of the world's economic requirements. Valuable reserves of phosphate used for fertilizers are mined in Idaho and adjacent states. Evaporite deposits in the interiors of continents, such as the potassium deposits near Carlsbad, New Mexico, indicate these areas were once inundated by ancient seas. Some limestone is chemically precipitated directly from seawater, and a minor amount precipitates in evaporite deposits from brines.

The most promising mineral deposits on the ocean floor are manganese nodules (Fig. 165). They are hydrogenous deposits, from Greek meaning "water generated." They form on the ocean floor by the slow accumulation of metallic elements extracted directly from seawater, which contains metals such as iron and manganese in solution at concentrations of less than one part per million by weight. The metals enter the oceans from streams that transport minerals derived from the weathering and decomposition of rocks on the continents and by hydrothermal vents on the ocean floor that acquire minerals from volcanically active zones beneath the crust.

Most metallic elements have a limited solubility in an alkaline, oxygen-rich environment such as seawater. Dissolved metals such as iron and manganese are oxidized by the presence of oxygen in seawater, forming insoluble oxides and hydroxides. The metals then deposit onto the ocean floor as tiny particles or as films or crusts covering any solid material on the seafloor. Living organisms also extract certain metals from seawater. When they die, their remains collect on the ocean floor, where the metals incorporate with the bottom sediments.

Most seafloor concretions such as manganese nodules are particularly well developed in deep, quiet waters far from continental margins and active volcanic spreading ridges. At these places, the steady rain of clay and other mineral particles prevents the metals from growing into concentrated deposits. The deposits occur in basins that receive a minimal inflow of sediments that would otherwise bury them. Such areas include abyssal plains and elevated areas on the ocean floor such as seamounts and isolated shallow banks.

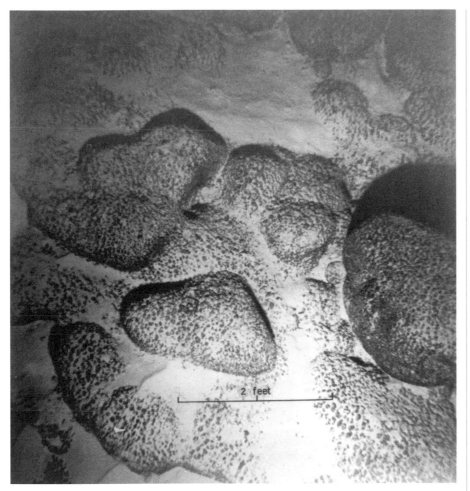

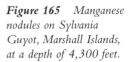

Figure 165 *Manganese nodules on Sylvania Guyot, Marshall Islands, at a depth of 4,300 feet.*

(Photo by K. O. Emery, courtesy USGS)

The manganese nodules grow around a solid nucleus, or seed, such as a grain of sand, a piece of shell, or a shark's tooth. The seed acts as a catalyst, enabling the metals to accrete to it like the growth of a pearl. Concentric layers accumulate until the nodules reach about the size of a potato, giving the ocean floor a cobblestone appearance. The growth rates of hydrogenous deposits are generally less than 1 inch in 10 million years.

A ton of manganese nodules contains about 600 pounds of manganese, 29 pounds of nickel, 26 pounds of copper, and about 7 pounds of cobalt. However, the location of these nodules at depths approaching 4 miles makes extraction on a large scale extremely difficult. About 100 square yards of bottom ooze must be sifted to extract a single ton of nodules. One mining method would use a dredge to scoop up the nodules. Another approach would employ a gigantic vacuum cleaner to suck up the nodules. A yet more exotic scheme envisions using television-guided robots to rake up the nodules, which are crushed into a slurry and pumped to the surface.

ENERGY FROM THE SEA

The world's oceans are a global solar collector. Daily, 30 million square miles of tropical seas absorb the equivalent heat content of 250 billion barrels of oil—greater than the world's total reserves of recoverable petroleum. If only a tiny fraction of this vast store of energy were converted into electricity, it could substantially enhance the world's future energy supply. The conversion of less than one-tenth of 1 percent of the heat energy stored in the surface waters of the tropics could generate roughly 15 million megawatts (million watts) of electricity, or more than 20 times the current generating capacity of the entire United States.

Ocean thermal-energy conversion, or OTEC (Fig. 166), takes advantage of the temperature difference between the surface and abyssal waters. Where a significant temperature difference exists between the warm surface water and the cold deep water, efficient electrical energy can be generated. In a

Figure 166 The ocean energy program at the National Renewable Energy Laboratory, Hawaii.

(Photo courtesy U.S. Department of Energy)

closed-cycle OTEC system, warm seawater evaporates a working fluid with a very low boiling point, such as Freon or ammonia. The working fluid enclosed in the system recycles continuously, similar to that in a refrigerator.

In an open-cycle OTEC system, also known as the Claude cycle after its inventor, the French biophysicist Georges Claude, the working fluid is a constantly changing supply of seawater. The warm seawater boils in a vacuum chamber, which dramatically lowers the boiling point. This system has the added benefit of producing desalinated water for irrigation in arid regions. In both systems, the resulting vapor drives a turbine to generate electricity. Cold water drawn up from depths of 2,000 to 3,000 feet condenses the gas back to a fluid to complete the cycle.

The nutrient-rich cold water could also be used for aquaculture, the commercial raising of fish, and serve nearby buildings with refrigeration and air-conditioning. The power plant could be located onshore, offshore, or on a mobile platform out to sea. The electricity could supply a utility grid system or be used on site to synthesize substitute fuels such as methanol and hydrogen, to refine metals brought up from the seabed, or to manufacture ammonia for fertilizer.

The open-cycle system offers several advantages over the closed-cycle system. By using seawater as the working fluid, the open-cycle system eliminates the possibility of contaminating the marine environment with toxic chemicals. The heat exchangers of an open-cycle system are cheaper and more effective than those used in a closed-cycle system. Therefore, open-cycle plants would more efficiently convert ocean heat into electricity and be less expensive to build.

Another source of energy is wave power. The breaking of a large wave on the coast is a vivid example of the sizable amount of energy that ocean waves produce. Waves are ultimately a form of solar energy. The Sun heats the surface of the ocean, producing winds that, in turn, drive the waves. Winds along the coast can also be harnessed to drive wind turbines (Fig. 167) to generate electricity. The best wave energy regions are generally along seacoasts at the receiving end of waves driven by the wind across large stretches of water. As the waves travel across the ocean, the winds continually pump energy into them. By the time they strike the coast, the waves have received a considerable amount of power.

The intertidal zones of rocky-weather coasts receive much more energy per unit area from waves than from the Sun. The waves form by strong winds from distant storms blowing across large areas of the open ocean. Local storms near the coasts provide the strongest waves, especially when superimposed on the rising and falling tides. Many hydroelectric schemes have been developed to utilize this abundant form of energy, which is economical and efficient. A crashing wave at the base of a wave-powered generator compresses the air at

Figure 167 *Wind turbines at San Gorgonio, California.*

(Photo courtesy U.S. Department of Energy)

the bottom of a chamber and forces it into a vertical tower, where the compressed air spins a turbine that drives an electrical generator.

Tidal power is another form of energy. Gulfs and embayments along the coast in most parts of the world have tides exceeding 12 feet, called macrotides. Such tides depend on the shapes of bays and estuaries, which channel the wavelike progression of the tides and increase their amplitude. The development of exceptionally high tidal ranges in certain embayments is due to the combination of convergence and resonance effects within the tidal basin. As the tide flows into a narrowing channel, the water movement constricts and augments the tide height.

Generating electricity using tidal power involves damming an embayment, letting it fill with water at high tide, and then closing the sluice gates at the tidal maximum when a sufficient head of water can drive the water turbines. Many locations with macrotides also experience strong tidal currents, which could be used to drive turbines that rotate with both the incoming and outgoing seawater to generate electricity.

Thermonuclear fusion energy (Fig. 168) is both renewable and essentially nonpolluting. The fuel for fusion is abundantly available in seawater. The energy from the fusion of deuterium, a heavy isotope of hydrogen, in a pool of water 100 feet on each side and 7 feet deep could provide the electrical needs of one-quarter of a million people for an entire year. Fusion is safe. Its by-products are energy and helium, a harmless gas that escapes into space.

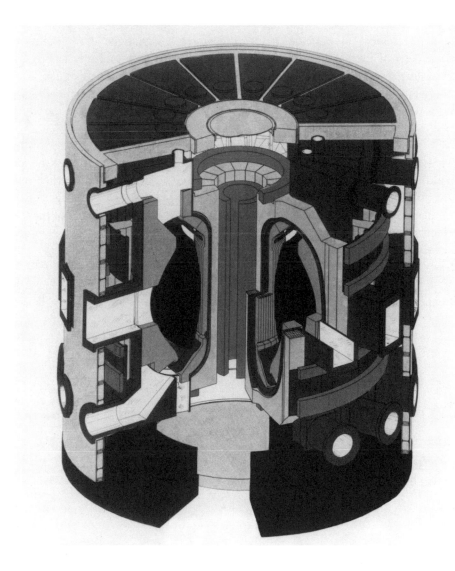

Figure 168 An artist's rendition of the International Fusion Experiment (ITER) at Princeton, New Jersey.

(Photo courtesy U.S. Department of Energy)

HARVESTING THE SEA

The world's fisheries are in danger of collapsing from overfishing. The United States created its marine sanctuaries program in 1972, when oil spills and treasure plundering began to pose a significant threat to its offshore resources. These sanctuaries prohibited oil drilling, salvaging, and other activities deemed harmful to the marine ecology. Yet all sanctuaries still allowed fishing. Most also permitted boating, mining, and other potentially disruptive activities. However, since the program's enactment, overfishing has become a much greater threat than oil pollution. Dwindling fish stocks such as cod and haddock have crashed in coastal waters, some to the brink of extinction.

The relative abundance of various species has changed dramatically in many parts of the world. The dangers result from a constant harvest rate of a dwindling resource caused by fluctuating environmental conditions, resulting in a major decline in fish catches. The composition of the catch is also changing toward smaller fish species. Even the average size of fish within the same species is becoming smaller.

Overfishing drives populations below levels needed for competition to regulate population densities of desired species. Therefore, under heavy exploitation, species that produce offspring quickly and copiously have a relative advantage. The extent to which these changes are due to shifts in fish populations, changes in patterns of commercial fishing, or environmental effects is uncertain. What is apparent is that if present trends continue, the world's fisheries could become smaller and composed of increasingly less desirable species.

The world's annual fish catch is about 100 million tons (Table 18), with the northwest Pacific and the northeast Atlantic yielding nearly half the

TABLE 18   PRODUCTIVITY OF THE OCEANS

Location	Primary Production Tons per Year of Organic Carbon	Percent	Total Available Fish Tons per Year of Fresh Fish	Percent
Oceanic	16.3 billion	81.5	0.16 million	0.07
Coastal Seas	3.6 billion	18.0	120.00 million	49.97
Upwelling Areas	0.1 billion	0.5	120.00 million	49.97
Total	20.0 billion		240.16 million	

total. A pronounced decline in heavily exploited fleshy fish are compensated by increased yields of so-called trash fish along with other small fish. The systematic removal of large predator fish might increase annual catches of other fish species by several million tons. However, such catches would consist of smaller fish that eventually dominate the northern latitudes, where population changes tend to be more variable and unpredictable than in the tropical regions.

Many changes in the world's fisheries are due to the strongly seasonal behavioral patterns of the fish as well as significant differences in climate and other environmental conditions from one season to the next. Climate influences fisheries by altering ocean surface temperatures, global circulation patterns, upwelling currents, salinity, pH balance, turbulence, storms, and the distribution of sea ice, all of which affect the primary production of the sea. Climatic conditions could cause a shift in species distribution and loss of species diversity and quantity.

To compensate for the shortfall in marine fisheries, a variety of aquatic animals are raised commercially for human consumption (Fig. 169). The shrimp, lobster, eel, and salmon raised by aquaculture account for less than 2 percent of the world's annual seafood harvest. However, their total value is estimated at five to 10 times greater than other fisheries. The development of aquaculture and mariculture could help meet the world's growing need for food. The Chinese lead the world with more than 25 million acres of impounded water in canals, ponds, reservoirs, and natural and artificial lakes that are stocked with fish.

The food requirements of the world might also be met by cultivating seaweed and algae, which are becoming important sources of nourishment rich in vitamins. The Japanese gather about 20 edible kinds of seaweed and consume weekly about 1 pound per person of dried algae preparations as appetizers or deserts, thereby becoming the world's leaders in the production of sea plants. The seaweed is harvested wild, and many varieties are also cultivated. When algae grows under controlled conditions, it multiplies rapidly and produces large quantities of plant material for food.

Algae crops can be harvested every few days, whereas agricultural crops grown on land require two to three months between planting and harvesting. An acre of seabed could yield 30 tons of algae a year compared with an average of 1 ton of wheat per acre of land. The algae can be artificially flavored to taste like meat or vegetables and is highly nutritious, containing more than 50 percent protein. The ocean farm is immensely rich and can meet human nutritional needs far into the future, provided people do not turn it into a desert as they have done with so much of the land.

After discussing the resources of the ocean, the next chapter will look at
the various types of creatures that live in the sea.

9

MARINE BIOLOGY
LIFE IN THE OCEAN

This chapter examines species living in the sea, including many unusual ones. Exploration of the ocean would not be complete without a view of its sea life. The riot of life in the tropical rain forests is repeated among the animals of the seafloor, especially the coral reef environment. The most primitive species, whose ancestors go back several hundred million years, anchor to the ocean floor.

Some of the strangest creatures on Earth live on the deep-ocean bottom. The seabed hosts an eerie world that time forgot. Tall chimneys spew hot, mineral-rich water that supports a variety of unusual animals in the cold, dark abyss. These unusual creatures have no counterparts anywhere else in the sea.

BIOLOGIC DIVERSITY

One of the most striking and consistent patterns of life on this planet is the greater the profusion of species when moving farther from the poles and closer to the equator. This is because near the equator, more solar energy is available for photosynthesis by simple organisms, the first link in the global food chain. Other factors that enter into this energy-species richness relationship include the climate, available living space, and the geologic history of the region. For instance, coral reefs and tropical rain forests support the largest species diversity because they occupy areas with the warmest climates.

The world's oceans have a higher level of species diversity than the continents. Due to a lower ecologic carrying capacity, which is the number of species an environment can support, the land has limited the total number of genera of animals since they first crawled out of the sea some 350 million years ago. The marine environment, by comparison, supports twice the living animal phyla than the terrestrial environment. Marine species have also existed twice as long as terrestrial species.

The oceans have far-reaching effects on the composition and distribution of marine life. Marine biologic diversity is influenced by ocean currents, temperature, the nature of seasonal fluctuations, the distribution of nutrients, the patterns of productivity, and many other factors of fundamental importance to living organisms. The vast majority of marine species live on continental shelves or shallow-water portions of islands and subsurface rises at depths less than 600 feet (Fig. 170). Shallow-water environments also tend to fluctuate more than habitats farther offshore, which affects evolutionary development. The richest shallow-water faunas live at low latitudes in the tropics, which are crowded with large numbers of highly specialized species.

When progressing to higher latitudes, diversity gradually falls off until reaching the polar regions, where less than one-tenth as many species live than

Figure 170 The distribution of shelf faunas.

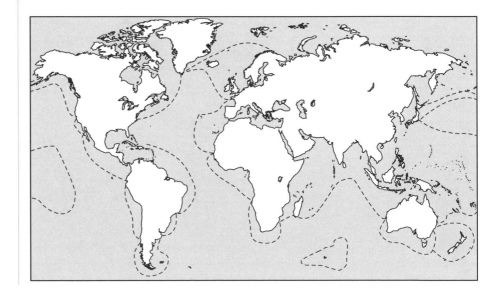

Figure 171 *Marine life on the bottom of McMurdo Sound, Antarctica.*

(Photo by W. R. Curtsinger, courtesy U.S. Navy)

in the tropics. Moreover, twice as much biologic diversity occurs in the Arctic Ocean, which is surrounded by continents, than in the Southern Ocean, which surrounds the continent of Antarctica. The sea around Antarctica is the coldest marine environment and was once though to be totally barren of life. Yet the waters around Antarctica are teeming with a large variety of species (Fig. 171). The Antarctic Sea represents about 10 percent of the total extent of the world's ocean and is the planet's largest coherent ecosystem. The abundance of species in the polar regions is due in most part to their ability to survive in subfreezing water.

The greatest biologic diversity is off the shores of small islands or small continents in large oceans, where fluctuations in nutrient supplies are least affected by the seasonal effects of landmasses. The least diversity is off large continents, particularly when they face small oceans, where shallow water seasonal variations are the greatest. Diversity also increases with distance from large continents.

Biologic diversity is highly dependent on the stability of food resources, which depend largely on the shape of the continents, the extent of inland seas, and the presence of coastal mountains. Erosion of mountains pumps nutrients into the sea, fueling booms of marine plankton and increasing the food supply for animals higher up the food chain. Organisms with abundant food are more likely to thrive and diversify into different species. Mountains that arise from the seafloor to form islands increase the likelihood of isolation of individual animals and, in turn, increase the chances of forming new species.

In the 1830s, when Charles Darwin visited the Galápagos Islands in the eastern Pacific (Fig. 172), he noticed major changes in plants and animals living on the islands compared with their relatives on the adjacent South American continent. Animals such as finches and iguanas assumed distinct but related forms compared with those on adjacent islands. Cool ocean currents and volcanic rock made the Galápagos a much different environment than Ecuador, the nearest land, which lies 600 miles to the east. The similarities among animals of the two regions could mean only that Ecuadorian species colonized the islands and then diverged by a natural process of evolution.

Continental platforms are particularly important because extensive shallow seas provide a large habitat area for shallow-water faunas and tend to dampen seasonal climatic variations, making the local environment more hospitable. As the seasons become more pronounced in the higher latitudes, food production fluctuates considerably more than in the lower latitudes. Species diversity is also influenced by seasonal changes such as variations in surface and upwelling ocean currents. These affect the nutrient supply and thereby cause large fluctuations in productivity.

Upwelling currents off the coasts of continents and near the equator are important sources of bottom nutrients such as nitrates, phosphates, and oxygen. Zones of cold, nutrient-rich upwelling water scattered around the world cover only about 1 percent of the ocean but account for about 40 percent of

Figure 172 Darwin's journey around the world during his epic exploration.

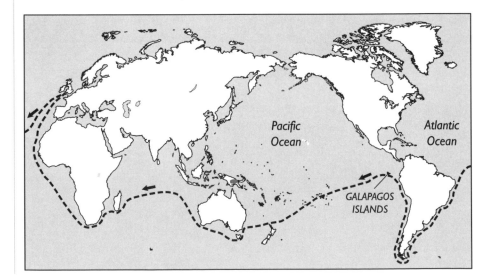

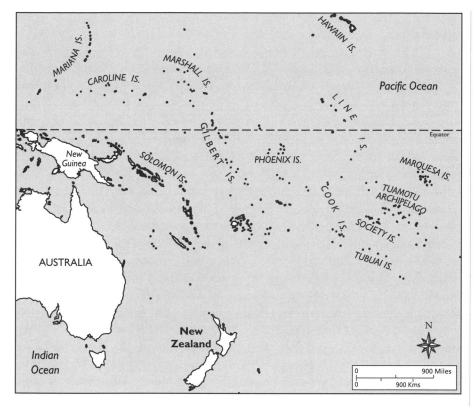

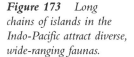

Figure 173 *Long chains of islands in the Indo-Pacific attract diverse, wide-ranging faunas.*

the ocean's productivity. These zones support prolific booms of phytoplankton and other marine life. These tiny organisms reside at the very bottom of the marine food web and are eaten by predators, which are preyed upon by progressively larger predators on up the food chain. These areas are also of vital economic importance to the commercial fishing industry.

Marine species living in different oceans or on opposite sides of the same ocean evolve separately from their overseas counterparts. Even along a continuous coastline, major changes in species occur that generally correspond to changes in climate. This is because latitudinal and climatic changes create barriers to shallow-water organisms. The great depth of the seafloor in some parts of the ocean provides another formidable barrier to the dispersal of shallow-water organisms. Furthermore, midocean ridges form a series of barriers to the migration of marine species.

These barriers partition marine faunas into more than 30 individual "provinces." Generally, only a few common species live in each province. The shallow-water marine faunas represent more than 10 times as many species than would be present in a world with only a single province. Such a condition existed some 200 million years ago when a single large continent was surrounded by a great ocean.

The Indo-Pacific province is the widest ranging of all marine provinces and the most diverse because of its long chains of volcanic island arcs (Fig. 173). When long island chains align east to west within the same climatic zone, they are inhabited by highly diverse, wide-ranging faunas. The faunas spill over from these areas onto adjacent tropical continental shelves and islands. However, this vast tropical biota is cut off from the western shores of the Americas by the East Pacific Rise, which is an effective obstruction to the migration of shallow-water organisms.

Biologic diversity mostly depends on the food supply. Small, simple organisms called phytoplankton (Fig. 174) are responsible for more than 95 percent of all marine photosynthesis. They play a critical role in the marine ecology, which spans 70 percent of Earth's surface. Phytoplankton are the primary producers in the ocean and occupy a key position in the marine food chain. They also produce 80 percent of the breathable oxygen as well as regulate carbon dioxide, which affects the world's climate.

The surface waters of the ocean vary markedly in color, depending on suspended matter such as phytoplankton, silt, and pollutants. In the open ocean, where the biomass is low, the water has a characteristic deep blue color. In the temperate coastal regions where the biomass is high, the water has a characteristic greenish color. The waters of the North Atlantic are colored green because they are richly endowed with phytoplankton.

Figure 174
Phytoplankton such as coccolithophores help maintain living conditions on Earth.

TABLE 19   CLASSIFICATION OF SPECIES

Group	Characteristics	Geologic Age
Vertebrates	Spinal column and internal skeleton. About 70,000 living species. Fish, amphibians, reptiles, birds, mammals.	Ordovician to recent
Echinoderms	Bottom dwellers with radial symmetry. About 5,000 living species. Starfish, sea cucumbers, sand dollars, crinoids.	Cambrian to recent
Arthropods	Largest phylum of living species with over 1 million known. Insects, spiders, shrimp, lobsters, crabs, trilobites.	Cambrian to recent
Annelids	Segmented body with well-developed internal organs. About 7,000 living species. Worms and leeches.	Cambrian to recent
Mollusks	Straight, curled, or two symmetrical shells. About 70,000 living species. Snails, clams, squids, ammonites.	Cambrian to recent
Brachiopods	Two asymmetrical shells. About 120 living species.	Cambrian to recent
Bryozoans	Moss animals. About 3,000 living species.	Ordovician to recent
Coelenterates	Tissues composed of three layers of cells. About 10,000 living species. Jellyfish, hydra, coral.	Cambrian to recent
Porifera	The sponges. About 3,000 living species.	Proterozoic to recent
Protozoans	Single-celled animals. Foraminifera and radiolarians.	Precambrian to recent

MARINE SPECIES

The most primitive of marine species are sponges (Table 19) of the phylum Porifera, which were the first multicellular animals. The sponge's body is composed of an outer layer and an inner layer of cells separated by a jellylike protoplasm. The cells can survive independently if separated from the main body. If a sponge is sliced up, individual pieces can grow into new sponges. The body walls of sponges are perforated by pores through which water is carried into the central cavity and expelled through one or more larger openings for feeding.

Certain sponge types have an internal skeleton of rigid, interlocking spicules composed of calcite or silica. One group has tiny glassy spikes for spicules, which give the exterior a rough texture unlike their softer relatives used in the bathtub. The so-called glass sponges consist of glasslike fibers of silica intricately arranged to form a beautiful network. The great success of the sponges and other organisms that extract silica from seawater to construct their skeletons explains why the ocean is largely depleted of this mineral. Some 10,000 species of sponges exist today.

The coelenterates, from Greek meaning "gut," include corals, hydras, sea anemones, sea pens, and jellyfish. They are among the most prolific of marine animals. No less than 10,000 species inhabit today's ocean. They have a saclike body with a mouth surrounded by tentacles. Most coelenterates are radially symmetrical, with body parts radiating outward from a central axis. Primitive, radially symmetrical animals have just two types of cells, the ectoderm and endoderm. In contrast, the bilaterally symmetrical animals also have a mesoderm (intermediate layer) and a distinct gut. During early cell division in bilateral animals, called cleavage, the fertilized egg forms two, then four cells, each of which gives rise to many small cells.

The corals come in large variety of forms (Fig. 175). Successive generations built thick limestone reefs. Corals began constructing reefs about 500 million years ago, forming chains of islands and barrier reefs along the shorelines of the continents. More recent corals are responsible for the construction of barrier reefs and atolls. They even rival humans in changing the face of the planet.

The coral polyp is a soft-bodied, contractible animal crowned with a ring of tentacles tipped with poisonous stingers that surround a mouthlike opening. The polyp lives in an individual skeletal cup, called a theca, composed of calcium carbonate. It extends its tentacles to feed at night and withdraws into the theca by day or during low tide to avoid drying out in the sun.

The corals live in symbiosis (living together) with zooxanthellae algae within their bodies. The algae ingest the corals' waste products and produce

Figure 175   *A collection of corals at Saipan, Mariana Islands.*

(Photo by P. E. Cloud, courtesy USGS)

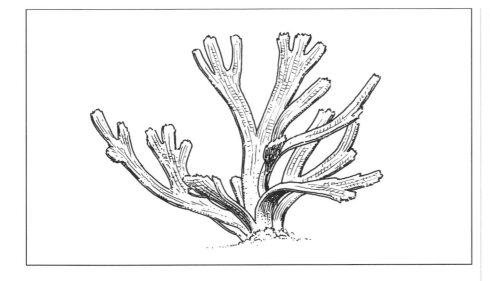

Figure 176   *The extinct bryozoans were major Paleozoic reef builders.*

nutrients that nourish the polyps. Since the algae need sunlight for photosynthesis, corals are restricted to warm ocean waters less than 300 feet deep. Much of the coral growth occurs within the intertidal zone. Widespread coral reef building occurs in warm, shallow seas with little temperature variation. Dense colonies of corals indicate conditions when the temperature, sea level, and climate are conducive to rapid coral growth.

The bryozoans (Fig. 176), or moss animals, are an unusual group of animals that live in extensive colonies attached to the seafloor. They filter feed on microscopic organisms. They are similar in appearance to corals but are more closely related to brachiopods. Bryozoan colonies show a considerable variety of forms, including branching, leaflike, and mosslike, giving the ocean floor a mossy appearance. Like corals, bryozoans are retractable animals encased in a calcareous vaselike structure, in which they retreat for safety. Bryozoans have simple calcareous skeletons in the shape of tiny tubes or boxes.

A new colony of bryozoans forms from a single, free-moving larval bryozoan that fixes onto a solid object and grows into numerous individuals by a process of budding, which is the production of outgrowths. The polyp has a circle of ciliated tentacles that form a sort of net around the mouth and are used for filtering microscopic food floating by. The tentacles rhythmically beat back and forth, producing water currents that aid in capturing food. Digestion occurs in a U-shaped gut. Wastes are expelled outside the tentacles just below the mouth.

The echinoderms, whose name means "spiny skin," are perhaps the strangest marine species. Their fivefold radial symmetry makes them unique among the more complex animals. They are the only animals possessing a water vascular system composed of internal canals that operate a series of tube feet or podia used for locomotion, feeding, and respiration. The great success of the echinoderms is illustrated by the fact that they have more classes of organisms than any phylum both living and extinct.

The major classes of living echinoderms include starfish, brittle stars, sea urchins, sea cucumbers, and crinoids. Sea cucumbers, named so because of their shape, have large tube feet modified into tentacles and a skeleton composed of isolated plates. The crinoids (Fig. 177), known as sea lilies because of their plantlike appearance, have long stalks composed of calcite disks, or columnals, anchored to the ocean floor by a rootlike appendage. Perched atop the stalk is a cup called a calyx that houses the digestive and reproductive systems. Up to 10,000 living species occupy the ocean depths.

The brachiopods, also called lampshells due to their likeness to primitive oil lamps, were once the most abundant and diverse marine organisms, with more than 30,000 species cataloged from the fossil record. Although plentiful during the Paleozoic, few living species are in existence. They are similar in appearance to clams and scallops, with two saucerlike shells fitted face-to-face that open and close using simple muscles. More advanced species called articulates have ribbed shells with interlocking teeth that maneuver along a hinge line.

Figure 177 *Crinoids grow upward of 10 feet or more tall.*

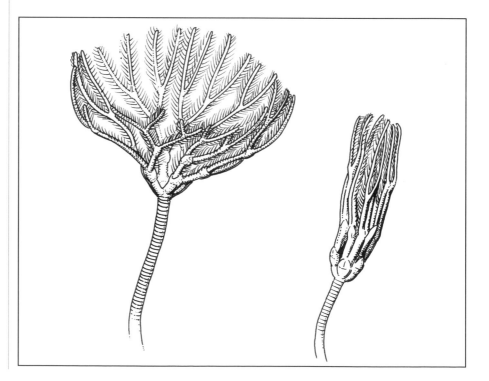

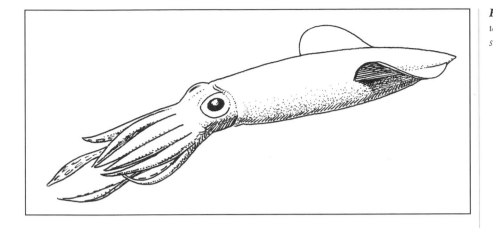

Figure 178 *The squids were among the most successful cephalopods.*

The shells are lined on the inside with a membrane called a mantle. It encloses a large central cavity that holds the lophophore, which functions in food gathering. Projecting from a hole in the valve is a muscular stalk called a pedicel by which the animal is attached to the seabed. The shells have a wide variety of forms, including ovoid, globular, hemispherical, flattened, convex-concave, and irregular. The surface is smooth or ornamented with ribs, grooves, or spines. The brachiopods filter food particles through opened shells that close to protect the animals against predators. Most modern brachiopods thrive in shallow waters or in intertidal zones. However, many inhabit the ocean bottom between 150 and 1,500 feet, with some thriving at depths reaching 18,000 feet.

The mollusks are a highly diverse group of marine animals and make up the second largest of the 21 animal phyla. Finding common features among various members is often difficult. The three major groups are the snails, clams, and cephalopods. The mollusk shell is an ever-growing one-piece coiled structure for most species and a two-part shell for clams and oysters. Mollusks have a large muscular foot for creeping or burrowing. Some have tentacles for seizing prey. Snails and slugs comprise the largest group.

The clams are generally burrowers, although many are attached to the ocean floor. The clam's shell consists of two valves that hang down on either side of the body and are mirror images of each other except in scallops and oysters. The cephalopods, which include the cuttlefish, octopus, nautilus, and squid (Fig. 178), travel by jet propulsion. They suck water into a cylindrical cavity through openings on each side of the head and expel it under pressure through a funnel-like appendage. As many as 70,000 species of mollusks inhabit the world today.

The nautilus (Fig. 179) is often referred to as a living fossil because it is the only extent relative of the swift-swimming ammonoids, which left a large variety of fossil shells. It lives in the great depths of the South Pacific and Indian Oceans down to 2,000 feet. The octopus, which also lives in deep waters, is somewhat like an alien life-form. It is the only animal with copper-based blood, whereas the blood of other animals is iron based.

The annelids are segmented worms, whose body is characterized by a repetition of similar parts in a long series. The group includes marine worms, earthworms, flatworms, and leeches. Marine worms burrow in the bottom sediments or are attached to the seabed, living in tubes composed of calcite or aragonite. The tubes are almost straight or irregularly winding and are attached to a solid object such as a rock, a shell, or coral. The prolific worms are represented by nearly 60,000 living species.

The arthropods are the largest group of marine and terrestrial invertebrates, comprising roughly 1 million species or about 80 percent of all known animals. The arthropods conquered land, sea, and air and are found in every environment on Earth. They include crustaceans, arachnids, and insects. The marine group includes shrimp, lobsters, barnacles, and crabs. The arthropod body is segmented, with paired, jointed limbs generally present on most segments and modified for sensing, feeding, walking, and reproduction. The body is covered with an exoskeleton composed of chitin that must be molted to accommodate growth. The crustaceans comprise about 40,000 living species.

Small shrimplike marine crustaceans called krill (Fig. 180) overwinter beneath the Antarctic ice, grazing off the ice algae. Krill serve as a major food source for other animals on up to whales. The biomass of krill exceeds that of

Figure 179 *The nautilus is the only living relative of the ammonoids.*

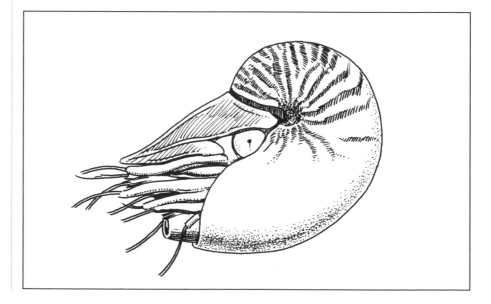

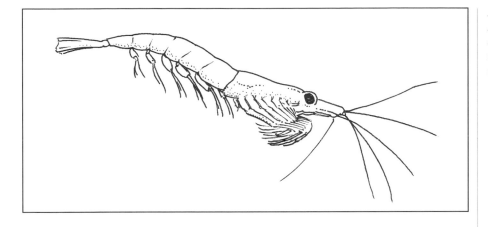

Figure 180   Krill are small, shrimplike crustaceans that are a major food resource for other marine species.

any other animal species, amounting to well over 1 billion tons. As whale populations decline, other krill-eating animals have shown rapid population increases in recent years, causing a subsequent decline in krill. Population increases in Antarctic seals have outpaced any simple recovery from past overhunting. Some seabird population increases have been documented as well. Similarly, populations of penguins (Fig. 181) are unexplainably larger after the slaughter of the 19th century. The penguin is one of the world's hardiest birds, able to nest along the harsh Antarctic coastline.

Figure 181   Strap penguins on ice floes in Arthur Harbor, Antarctica.

(Photo by G.V. Graves, courtesy U.S. Navy)

Fish comprise more than half the species of vertebrates. They include the jawless fish (lampreys and hagfish), the cartilaginous fish (sharks, skates, rays, and ratfish), and the bony fish (salmon, swordfish, pickerel, and bass). The ray-finned fish are by far the largest group of fish species. Sharks have been highly successful for the last 400 million years. They play a critical role by preying on sick and injured fish and thus help keep the ocean healthy. Closely related to the sharks are the rays (Fig. 182), whose pectoral fins are enlarged into wings, allowing them literally to fly through the sea. Today, fish comprise about 22,000 species.

Marine mammals called cetaceans include whales, porpoises, and dolphins, all of which evolved during the last 50 million years. Sea otters, seals, walruses, and manatees (Fig. 183) are not fully adapted to a continuous life at sea and have retained many of their terrestrial characteristics. Whales have adapted to swimming, diving, and feeding that matches or surpasses fish and sharks. They might have gone through a seal-like amphibious stage early in their evolution. Today, their closest relatives are the artiodactyls, or hoofed mammals with an even number of toes, such as cows, pigs, deer, camels, and giraffes. The giant blue whale (Fig. 184) is the largest animal on Earth, even dwarfing the biggest dinosaurs that ever lived.

The pinnipeds, meaning "fin-footed," are a group of marine mammals with four flippers whose three surviving forms include seals, sea lions, and walruses. The "true" seals without ears are thought to have evolved from weasel-like or otterlike forms, whereas sea lions and walruses are believed to have developed from bearlike forms. The similarity in their flippers, however,

Figure 182 *The rays fly through the sea on extended pectoral fins.*

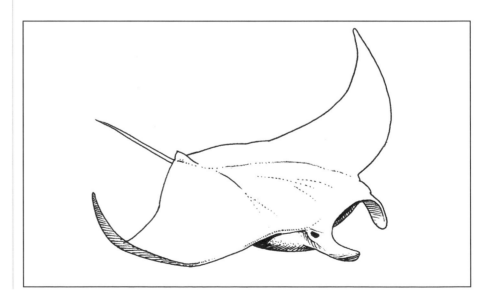

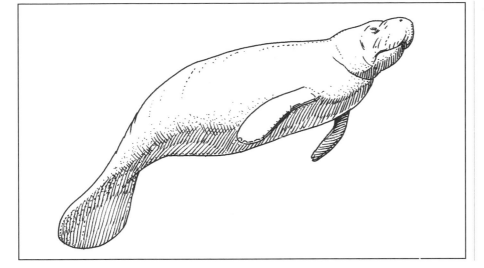

Figure 183 *Manatees are threatened with extinction.*

Figure 184 *The blue whale* (bottom) *is the largest mammal on Earth.*

suggests that all pinnipeds evolved from a single land-based mammal that entered the sea millions of years ago.

LIFE IN THE ABYSS

The deep waters of the open ocean were once thought to be a lifeless desert. While dredging the ocean bottom in the 1870s, the British oceanographic ship *Challenger* hauled to the surface a large collection of deepwater and bottom-dwelling animals from even the deepest trenches, including hundreds of species never seen before and unknown to science. The catch comprised some of the most bizarre life-forms molded by adaptive behavior and natural selection to the cold and dark of the abyss along with several species thought to have long been extinct.

A century later, a population of large, active animals were discovered thriving in total darkness as deep as 4 miles. These depths were previously thought to be the domain of small, feeble creatures such as sponges, worms, and snails that were specially adapted to live off the debris of dead animals raining down from above. In fact, much of the deep seafloor was later found to be teaming with many species of scavengers, including highly aggressive worms, large crustaceans, deep-diving octopuses, and a variety of fish including giant sharks.

The large physical size of many species is due to an abundance of food, lower levels of competition, and lack of juveniles, which live in the shallower depths and descend to deeper water when they mature. Large numbers of fish from the great depths of the lower latitudes are related to shallow-water varieties of the higher latitudes. Some Arctic fish might represent near-surface expressions of populations that inhabit the cold, deep waters off continental margins.

The coelacanth (Fig. 185), once thought to have gone extinct along with the dinosaurs and ammonoids, stunned the scientific world in 1938 when fishermen caught a 5-foot specimen in the deep, cold waters of the Indian Ocean off the Comoro Islands near Madagascar. The discovery caused much sensation for at last here was the missing link between fish and tetrapods. The fish looked ancient, a castaway from the distant past. It had a fleshy tail, a large set of forward fins behind the gills, powerful square toothy jaws, and heavily armored scales. Stout fins enabled the fish to crawl along on the deep ocean floor in much the same manner its ancestors crawled out of the sea to populate the land.

The oldest species living in the world's oceans thrive in cold waters. Many Arctic marine animals, including certain brachiopods, starfish, and bivalves, belong to biologic orders whose origins extend back hundreds of millions of years. Some 70 species of marine mammals, including dolphins,

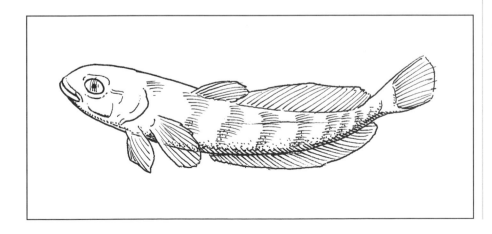

Figure 185 The
coelacanth lives in the
deep waters of the Indian
Ocean.

porpoises, and whales, spend much of their time feeding in the cold Arctic
waters of the polar regions.

The Antarctic Sea is the coldest marine habitat in the world. It was once
though to be totally barren of life. However, in 1899, a British expedition to
the southernmost continent found examples of previously unknown fish
species related to the perches (Fig. 186) common to many parts of the world.
Upward of 100 species of fish are confined to the Antarctic region, account-
ing for about two-thirds of the fish species in the area. Because they live in
subfreezing waters, the fish rely on special blood proteins that bind to ice crystals. This keeps the ice crystals from growing, somewhat like antifreeze to help
the fish to survive during the cold Antarctic winters.

A circum-Antarctic current isolates the Antarctic Sea from the general
circulation of the ocean and serves as a thermal barrier. It impedes the inflow

Figure 186 Antarctic
perchlike fish produce
antifreeze-like substances
to keep from freezing.

245

of warm currents and warm-water fish as well as the outflow of Antarctic fish. Also, due to the extreme cold and low productivity, the Antarctic Sea is less diverse than the Arctic Ocean, which supports almost twice as many species.

In the great depths of the abyss, animals live in the cold and dark, adapting to such high pressures that they perish when brought to the surface. Some bacteria along with higher life-forms live successfully at extreme pressures of more than 1,000 atmospheres (atmospheric pressure at sea level) in the deepest parts of the sea but cannot grow at pressures of less than 300 atmospheres. The bacteria, which account for half of all organic carbon in the oceans, aid in the decay of dead plant and animal material that falls to the deep seafloor to recycle organic matter in the sea.

On the floor of the Gulf of Mexico 1,800 feet below the surface waters, marine biologists discovered in the cold and dark what appears to be a new, remarkable worm species living among mounds of frozen natural gas called methane hydrate that seeped from beneath the seabed. This crystallized blend of water ice and natural gas forms a rocklike mass in great abundance in the high pressures and low temperatures of the deep sea. Estimates indicate that enough methane is locked within hydrates to blanket the entire Earth with a layer of gas 160 feet thick.

The Gulf of Mexico is one of the few places where lumps of methane hydrate actually break through the seafloor. The worms are flat, pink centipede-like creatures 1 to 2 inches long. They live in dense colonies that tunnel through the 6-foot-wide ice mounds shaped like mushrooms on the ocean floor. The worms appear to survive by eating bacteria growing on the yellow and white ice mounds. They might also feed directly on the methane within the hydrates. Despite their rock-solid appearance, the hydrates are generally unstable. Slight changes in temperature or pressure caused by the tunneling worms could cause the methane ice to melt, prompting the seafloor above it to collapse to the detriment of the worms.

CORAL REEFS

Coral reefs are the oldest ecosystems and important land builders in the tropics, forming entire chains of islands and altering the shorelines of the continents (Fig 187). Over geologic time, corals have built massive reefs of limestone. The reefs are limited to clear, warm, sunlit tropical oceans such as the Indo-Pacific and the western Atlantic. Hundreds of atolls comprising rings of coral islands that enclose a central lagoon dot the Pacific. The islands consist of reefs several thousand feet across, many of which formed on ancient volcanic cones that have vanished below the waves, with the rate of coral growth matching the rate of subsidence.

The coral-reef environment supports more plant and animal species than any other habitat. The key to this prodigious growth is the unique biology of corals, which plays a vital role in the structure, ecology, and nutrient cycles of the reef community. Coral reef environments have among the highest rates of photosynthesis, nitrogen fixation, and limestone deposition of any ecosystem. The most remarkable feature of coral colonies is their ability to build massive calcareous skeletons weighing several hundred tons.

Coral reefs forming in shallow water where sunlight can easily penetrate for photosynthesis contain abundant organic material. More than 90 percent of a typical reef consists of fine, sandy detritus stabilized by plants and animals anchored to the reef surface. Tropical plant and animal communities thrived on the reefs due to the corals' ability to build massive, wave-resistant structures. The major structural feature of a living reef is the coral rampart that reaches almost to the water's surface. It consists of massive rounded coral heads and a variety of branching corals (Fig. 188).

Living on this framework are smaller, more fragile corals and large communities of green and red calcareous algae. Hundreds of species of encrusting organisms such as barnacles thrive on the coral framework. Large

Figure 187 *A fringing coral reef in Puerto Rico.*

(Photo by C. A. Kaye, courtesy USGS)

247

numbers of invertebrates and fish hide in the nooks and crannies of the reef, often waiting until night before emerging to feed. Other organisms attach to virtually all available space on the underside of the coral platform or onto dead coral skeletons. Filter feeders such as sponges and sea fans occupy the deeper regions.

Fringing reefs grow in shallow seas and hug the coastline or are separated from the shore by a narrow stretch of water. Barrier reefs also parallel the coast but are farther out to sea, are much larger, and extend for longer distances. The best example is the Great Barrier Reef (Fig. 189), a chain of more than 2,500 coral reefs and small islands off the northeastern coast of Australia. It forms an underwater embankment more than 1,200 miles long, up to 90 miles wide, and as much as 400 feet above the ocean floor. It is one of the great wonders of the world and the largest feature built by living organisms. The Great Barrier Reef is a relatively young structure. It formed largely during the Pleistocene ice ages when sea levels fluctuated with the growth of continental glaciers over the last 3 million years.

The second largest reef is the Belize Barrier Reef complex on the Caribbean coast of South America. It is the most luxuriant array of reefs in the Western Hemisphere. Extensive reefs also rim the Bahama Banks archipelago. A small reef that fringes Costa Rica is in danger from pollution from pesticides and soil runoff. Other reefs throughout the world are similarly affected.

Figure 188 Coral at Bikini Atoll, Marshall Islands.

(Photo by K. O. Emery, courtesy USGS)

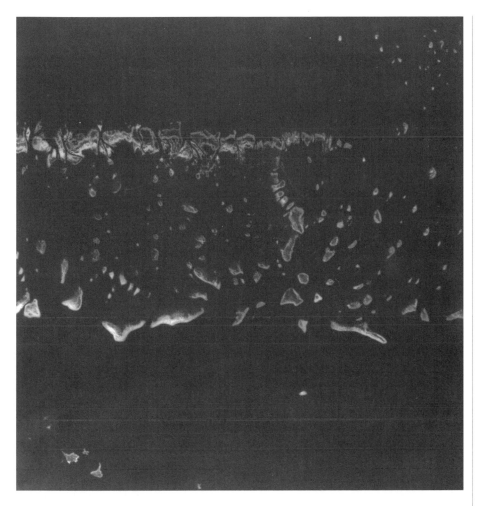

Figure 189 *The Great Barrier Reef northwest of Australia is the world's most extensive reef system.*

(Photo courtesy NASA)

The structure of the reef includes the fore reef, which is seaward of the reef crest, where coral blankets nearly the entire seafloor. In deeper waters, many corals grow in flat, thin sheets to maximize their light-gathering area. In other parts of the reef, the corals form massive buttresses separated by narrow sandy channels composed of calcareous debris from dead corals, calcareous algae, and other organisms living on the coral. The channels resemble narrow winding canyons with vertical walls of solid coral. They dissipate wave energy, allowing the free flow of sediments to prevent the coral from choking on the debris.

Below the fore reef is a coral terrace, followed by a sandy slope with isolated coral pinnacles, then another terrace, and finally a nearly vertical drop-off into the dark abyss. The rise and fall of sea levels during the last few million years have produced terraces that resemble a stair-step growth of coral run-ning up an island or a continent. The coral skeletons represent periods of extensive glaciation when sea levels dropped as much as 400 feet. In Jamaica, almost 30 feet of reef have built up since the present sea level stabilized some 5,000 years ago following the last ice age.

Coral reefs are centers of high biologic productivity where fish and other creatures hide among the crooks and crannies in the coral (Fig. 190). They also sustain fisheries that are a major source of food in the tropical regions. Unfortunately, the spread of tourist resorts along coral coasts in many parts of the world is harming the productivity of these areas due mostly to an increase in sedimentation. These developments are usually accompanied by increased sewage dumping, overfishing, and physical damage to the reef from construction, dredging, dumping, and landfills along with the direct destruc-tion of the reef for souvenirs and curios.

In areas such as Bermuda, the Virgin Islands, and Hawaii, development and sewage outflows have caused extensive overgrowth and death of the reef by thick mats of algae (not to be confused with algae within the coral itself). The algae suffocate the coral by supporting the growth of oxygen-consuming bacteria, particularly in the winter, when the algal cover on shallow reefs is high. This action results in the death of the living coral and the eventual destruction of the reef and the biologic communities it supports.

Figure 190 *A species of angelfish swims among the rock and coral off Andros Island, Bahamas.*

(Photo by P. Whitmore, courtesy U.S. Navy)

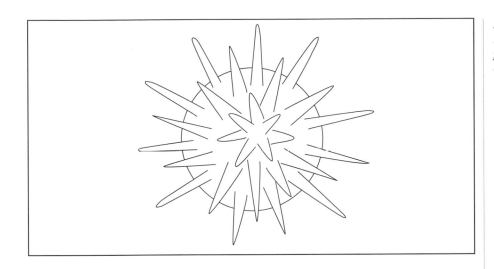

Figure 191
Foraminiferans are tiny
planktonic marine
organisms.

The increased warming of the world's ocean is bleaching many reefs, causing the corals to turn ghostly white due to the expulsion of algae from their tissues. The algae aid in nourishing the corals, and their loss poses a great danger to the reefs. Foraminifers (Fig. 191), which are tiny marine organisms whose shells often are perforated with minute holes for the protrusion of slender pseudopodia, exhibit damage similar to that observed in bleached coral reefs. Along with corals, they play an important role in the global ecosystem wherein bleaching could seriously affect the marine food chain.

THE VENT CREATURES

On the crest of the East Pacific Rise south of Baja California, 8,000 feet beneath the ocean lies an eerie world of unexpected beauty. In volcanically active fields, an oasis of hydrothermal vents and exotic deep-sea creatures previously unknown to science, including blind shrimps, unusual echinoderms, and metal-eating bacteria, thrive in the cold, dark abyss. The base of jagged basalt cliffs display evidence of active lava flows, including fields strewn with pillow lavas as though extruded from a giant toothpaste tube. Perched near the seams of midocean ridges, curious pillars of lava stand like Greek columns 45 feet tall.

Undersea geysers build forests of exotic chimneys called black smokers that spew out hot water blackened with sulfur compounds. The largest known black smoker is a 160-foot-tall structure off the coast of Oregon. The black smokers support some of the world's most bizarre biology. Flourishing among the hydrothermal vents are perhaps the strangest animals ever encountered due to their unusual habitat. Life might even have originated around such vents, obtaining from Earth's hot interior all the nutrients necessary for survival.

Seawater seeping downward near magma chambers acquires heat and minerals and is expelled through fissures in the ocean floor. The hydrothermal vents not only maintain the bottom waters at livable temperatures, as much as 20 degrees Celsius, they also provide valuable nutrients. This makes the environment totally independent of the Sun as a source of energy, which instead comes from Earth itself.

Clustered around the hydrothermal vents are large communities of unusual animals (Fig. 192) as crowded as any tropical rain forests. Large white clams up to 1 foot long and mussels, having no need for skin pigments, nestle between black pillow lava. Giant white crabs scamper blindly across the volcanic terrain, and long-legged marine spiders roam the ocean floor. They live in total darkness and therefore have no need for eyes, which have become useless appendages. More than 300 new species of vent animals have been identified since the first hydrothermal vent was discovered in 1977.

The most bizarre of the vent creatures are clusters of giant tube worms up to 10 feet tall that sway in the hydrothermal currents. The tube worms are

Figure 192   Tall tube worms, giant clams, and large crabs live on the deep-sea floor near hydrothermal vents.

contractible animals residing inside long, white stalks up to 4 inches wide. They can increase in length at a rate of more than 33 inches per year, making them the fastest growing marine invertebrates. While feeding, tube worms expose a long, bright, gill-like, red plume abundantly supplied with blood. They use these plumes to collect hydrogen sulfide, nitrate, and other nutrients to feed symbiotic bacteria living in their guts. The bacteria break down these compounds to provide nutrition for the worms.

The brown, spongy tissue filling the inside of tubeworms is packed with bacteria, comprising some 300 billion organisms per ounce of tissue. The plume is also a delicacy for hungry crabs, which attempt to climb the stalks in search of a meal. Tubeworms reproduce by spawning, releasing sperm and eggs. These combine in seawater to create a new worm complete with its own starter set of bacteria. The worms are so accustomed to the high pressure at these great depths that they immediately perish and come out of their tubes when brought to the surface.

In the Atlantic, the vents are dominated by swarms of small shrimps. They were originally thought to be blind until the discovery of an unusual pair of vision organs on the shrimps' backs to replace the eyes attached to their heads. Apparently, they see by the feeble light emanating from the hot-water chimneys. Thermal radiation alone cannot explain the light emitted by the 350-degree Celsius water. The light is possibly produced by the sudden cooling of the hot water, which produces crystalloluminescence as dissolved minerals crystallize and drop out of solution, emitting light in the process. The light, although extremely dim, is apparently sufficient to allow photosynthesis to take place even on the very bottom of the ocean. The deep-sea light has intrigued biologists because of the possibility that organisms can harness this energy by using a type of photosynthesis totally independent of the Sun.

Nearly every food web on the planet is supported by photosynthetic organisms using solar energy to produce carbohydrates. In 1984, during an expedition using the deep-diving submersible *Alvin,* researchers discovered groups of animals thriving in the extreme depths of the Gulf of Mexico, well beyond the reach of the Sun's rays and far from seafloor hot springs. Communities of mussels, tubeworms, crabs, fish, and other marine animals thrive on the cold seafloor at the base of the Florida escarpment. The animals live in symbiosis with bacteria, which graze off organic matter disseminated in the porous rock.

The most remarkable aspect about the vent animals is that they do not obtain their nutrition in the form of detrital material falling from above. Instead, they rely on symbiosis with sulfur-metabolizing bacteria that live within the host's tissues. The bacteria metabolize sulfur compounds in the hydrothermal water by chemosynthesis. They harness energy liberated by the oxidation of hydrogen sulfide from the vents to incorporate carbon dioxide for the production of organic compounds such as carbohydrates, proteins, and lipids. The by-products of the bacteria's metabolism are absorbed into the host animal and nourish it. The vent animals are so dependent on the bacteria that the mussels have only a rudimentary stomach and the tubeworms lack even a mouth.

Some animals also feed on bacteria directly. Odd-looking colonies of bacteria, some with long tendrils that sway in the warm currents, become the feeding grounds for complex forms of higher life. Some regions are clouded by drifting bits of whitish bacteria that swirl like falling snow. Occasionally, clumps of waving bacteria break loose from fissures and join the swirl of biologic snowfall.

The animals live precarious lives since the hydrothermal vent systems turn on and off sporadically. Species can survive only as long as the vents continue to operate—perhaps on timescales of only a few years. To illustrate this point, isolated piles of empty clamshells bear witness to local mass fatalities. In new basalt fields, the vent creatures soon establish residency around young hydrothermal vents. Then the once-barren abyss is rapidly colonized.

THE INTERTIDAL ZONE

The constant waxing and waning of the tides are responsible for the prodigious growth in the intertidal zone (Fig. 193), the habitat area between high

Figure 193 Intertidal exposure of chaotic blocks of sandstone near Clallam County, Washington.

(Photo by W. O. Addicott USGS)

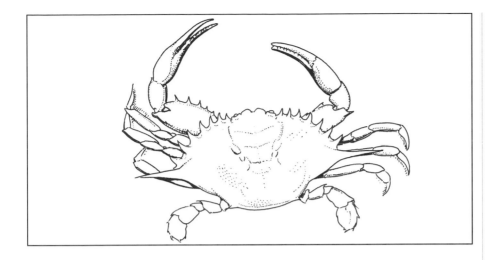

Figure 194 *The activity of crabs is attuned to the tidal rhythms.*

and low tides. Most inhabitants of the intertidal zone have biologic clocks set to the rhythm of the lunar day. The rhythms are characterized by repetitive behavioral or physiological events such as feeding and resting that are synchronized with the tides. Each lunar day, which is about 25 hours long, generally has two tides. This produces bimodal lunar-day rhythms, as compared with the unimodal solar-day rhythms of organisms attuned to the 24-hour solar day.

Biologic clocks are important aids to survival by giving advance warning of regular changes in the environment such as nightfall or the return of the tides. Even under constant laboratory conditions absent the effects of diurnal or tidal cycles, biologic clocks continue to function and the tidal rhythms persist for some time.

Apparently, the tidal rhythms are not learned or impressed on the organisms by the tides themselves. Crabs (Fig. 194) are among the most active animals of the intertidal zones. Crabs raised in the laboratory and exposed only to diurnal conditions exhibit a distinct tidal component in their activity when their body temperatures are lowered. Also, crabs taken from areas not subject to tides and moved to a tidal flat quickly establish a tidal rhythm. Apparently, the clock that measures the tidal frequency is innate and needs only to be activated by an outside stimulus.

Rhythmic behavior in organisms is also an expression of the genetic code. Heredity decides whether an animal is active during high or low tide. The environment also plays an important role in the establishment of a tidal rhythm. The schedule of the tides determines only the setting of the biologic clock. Therefore, animals transported to a different ocean synchronize their clocks to the new tidal conditions. Moreover, the pounding surf shapes the activity patterns of inhabitants living on beaches exposed to the open sea.

Intertidal organisms living in protected bays are not as exposed to the vicissitudes of the sea. Instead, they are controlled by more subtle conditions such as a drop in temperature or pressure changes induced by the incoming tides, which help set their tidal rhythms. Even without outside stimuli, the biologic clock continues to run accurately but no longer controls the organism's activity. It operates independently from tidal influences until the organism returns to the sea and the clock takes over again. As with all clocks, the accuracy of biologic clocks is not altered by changes in the environment nor do they pertain to intertidal organisms alone. Instead, they affect the entire spectrum of life.

After discussing marine species, the final chapter explores some of the strangest geologic features of the ocean floor.

10

RARE SEAFLOOR FORMATIONS

UNUSUAL GEOLOGY ON THE SEABED

This chapter examines many unique geologic features on the ocean floor. The sea is often reluctant to give up its secrets, and many unexplained wonders still persist (Fig. 195). The floor of the ocean is host to a myriad of exotic geologic structures found nowhere else on Earth. Instead of hot lava as with most volcanoes, unusual seamounts associated with deep-sea trenches erupt cold mud composed of plastic serpentine. Scattered around the midocean ridges are remarkable volcanic deposits, including piles of pillow lavas, forests of black smokers, and undersea geysers that eject vast quantities of hot water that rise toward the surface in massive plumes.

Submarine slides larger than any landslide gouge out deep chasms in the ocean floor and deposit enormous heaps of sediment onto the bottom of the sea. The active seafloor sports a variety of depressions, including sea caves, blue holes, gas blowouts, calderas, and numerous craters formed by undersea explosions. Large meteorites or comets falling into the sea blast deep craters in the ocean floor, many of which are better preserved than their landward counterparts.

Figure 195 *An unusual lightning strike of a plume of water in the ocean.*

(Photo courtesy U.S. Navy)

MUD VOLCANOES

In the western Pacific Ocean, about 50 miles west of the Mariana Trench, the world's deepest depression, lies a cluster of large seamounts 2.5 miles below the surface of the sea in a zone about 600 miles long and 60 miles wide. The undersea mountains were built not by hot volcanic rock as with most Pacific seamounts but by cold serpentine, which is a soft, mottled green rock similar to the color of a serpent, hence its name. Serpentine is a low-grade metamorphic rock and the main mineral of asbestos. It originates from the reaction of water with olivine, an olive-green, iron- and magnesium-rich silicate that is a major constituent of the upper mantle.

The erupting serpentine rock flows down the flanks of the seamounts similar to lava from a volcano and forms gently sloping structures. Many of these seamounts rise more than 1 mile above the ocean floor and measure as much as 20 miles across at the base, resembling broad shield volcanoes such Mauna Loa (Fig. 196), which built the main island of Hawaii. Drill cores taken during the international Ocean Drilling Program in 1989 showed that serpentine not only covers the tops of the seamounts but also fills the interiors.

Several smaller seamounts only a few hundred feet high are mud volcanoes, resembling those in hydrothermal areas on land (Fig. 197). They are

Figure 196 *The Mauna Loa Volcano, Hawaii.*

(Photo courtesy USGS)

Figure 197 Mud
volcanoes and acidulated
ponds northwest of
Imperial Junction,
Imperial County,
California.

(Photo by Mendenhall,
courtesy USGS)

composed of mounds of remobilized sediments formed in association with hydrocarbon seeps, where petroleum-like substances ooze out of the ocean floor. Apparently, sediments rich in planktonic carbon are "cracked" into hydrocarbons by the heat of Earth's interior. Even drill cores recovered around hydrothermal fields smell strongly of diesel fuel.

Mud volcanoes exist in many places around the world. They usually develop above rising blobs of salt or near ocean trenches. The mud comprises peridotite that is converted into serpentine and ground down into rock flour called fault gouge by movement along underlying faults. The mud volcanoes appear to undergo pulses of activity interspersed with long dormant periods. Many seamounts formed recently (in geologic parlance), probably within the last million years or so.

A strange mud volcano that spews out a slurry of seafloor sediments mixed with water lies beneath the chilly waters of the Arctic Ocean. It is a half-mile-wide circular feature that lies 4,000 feet deep and is covered by an unusual layer of snowlike natural gas called methane hydrate. The underwater volcanic structure is the first of its kind found covered with such an icy coating draped across a warm mud volcano. Methane hydrate is a solid mass formed when high pressures and low temperatures squeeze water molecules into a crystalline cage around a methane molecule. Vast deposits of methane hydrate are thought to be buried in the ocean floor around the continents and represent the largest untapped source of fossil fuel left on Earth.

The Mariana seamounts appear to be diapirs similar to salt diapirs of the Gulf of Mexico, which trap oil and gas. The diapirs appear to be composed of the mantle rock peridotite altered by interaction with fluids distilled from the subducted portion of the Pacific plate as it descends into the Mariana Trench and slides under the Philippine plate. Fluids expelled from the subducting plate react with the mantle rock, transforming portions of the mantle into low-density minerals that rise slowly through the subduction zone to the seafloor.

About 90 million years ago, the Mariana region forward of the island arc consisted of midocean ridge and island arc basalts that have been eroding away as much as 40 miles by plate subduction over the last 50 million years. The seamount-forming process has been proceeding for perhaps 45 million years as oceanic lithosphere vanishes into the subduction zone, distilling enormous quantities of fluids from the descending plate. The fluids reacting with the surrounding mantle produce blobs of serpentine that rise to the surface through fractures in the ocean floor.

The fluid temperatures in subduction zones are cool compared with those associated with midocean ridges, where hydrothermal vents eject high-temperature black effluent. Instead of comprising heavy minerals like the black smokers of the East Pacific Rise and other midocean ridges, the ghostly white chimneys of the Mariana seamounts in the western Pacific near the world's deepest trench are composed of a form of aragonite. The rock is composed of white calcium carbonate with a very unusual texture that normally dissolves in seawater at these great depths. Hundreds of corroded and dead carbonate chimneys were found strewn across the ocean floor in wide "graveyards of eerie towers."

Apparently, cool water emanating from beneath the surface allows the carbonate chimneys to grow and avoid dissolution by seawater. Many carbonate chimneys are thin and generally less than 6 feet high. Other chimney structures are thicker, are taller, and occasionally coalesce to form ramparts encrusted with black manganese deposits. Small manganese nodules are also scattered atop many of the mountains of mud.

Exotic terranes are fragments of oceanic lithosphere originating from distant sources and exposed on the continents and islands in zones where plates collide. Many terranes contain large serpentine bodies that are similar in structure to the Mariana seamounts. Their presence is a constant reminder that the ocean floor was highly dynamic in the past and continues to be so today.

Tufa is a porous rock composed of calcite or silica that commonly occurs as an incrustation around the mouths of hot springs. However, in southwestern Greenland, more than 500 giant towers of tufa cluster together in the chilly waters of Ikka Fjord. Some reach as high as 60 feet, and their tops are visible at low tide. The towers are made of an unusual form of calcium carbonate called ikaite. Its crystals form when carbonate-rich water from springs beneath the fjord seeps upward and comes into contact with cold, calcium-laden seawater. Because of the low temperature, the water cannot escape during the precipitation of the mineral and is incorporated into the crystal lattice, producing weird, yet beautiful formations.

SUBSEA GEYSERS

Perhaps the strangest environment on Earth lies on the ocean floor in deep water near seafloor spreading centers such as the crests of the East Pacific Rise and the Mid-Atlantic Ridge, which are portions of Earth's largest volcanic system. Solidified lava lakes hundreds of feet long and up to 20 or more feet deep probably formed by rapid outpourings of lava. In places, the surface of a lava lake has caved in, forming a collapsed pit (Fig. 198).

Seafloor spreading is often described as a wound that never heals as magma slowly oozes out of the mantle in response to diverging lithospheric plates. During seafloor spreading, magma rising out of the mantle solidifies on the ocean floor, producing new oceanic crust. At the base of jagged basalt cliffs is evidence of active lava flows and fields strewn with pillow formations formed when molten rock ejects from fractures in the crust and is quickly cooled by the deep, cold water.

Figure 198 The rim of a lava lake collapse pit on the Juan de Fuca ridge in the East Pacific.

(Photo courtesy USGS)

Figure 199 *Pillow lava
on the ocean floor.*

(Photo courtesy WHOI)

Lava erupting from undersea volcanoes constantly forms new crust
along the midocean ridges as lithospheric plates on the sides of the rift inch
apart and molten basalt from the mantle slowly rises to fill the gap. Occasionally, a colossal eruption of lava along the ridge crest flows downslope for more
than 10 miles. Most of the time, however, the basalt just oozes out of the
spreading ridges, forming a variety of lava structures on the ocean floor.

The ridge system exhibits many uncommon features, including massive
peaks, sawtoothlike ridges, earthquake-fractured cliffs, deep valleys, and a
large variety of lava formations. Lava formations associated with midocean
ridges consist of sheet flows and pillow, or tube, flows. Sheet flows are more
common in the active volcanic zone of fast spreading ridge segments such as
those of the East Pacific Rise, where the plates are separating at a rate of 4 to
6 inches a year.

Pillow lavas (Fig. 199) erupt as though basalt were squeezed out onto
the ocean floor. They generally arise from slow spreading centers, such as
those of the Mid-Atlantic Ridge. There plates spread apart at a rate of only
about 1 inch per year, and the lava is much more viscous. The surface of the
pillows often contains corrugations or small ridges, indicating the direction
of flow. The pillow lavas typically form small, elongated hillocks pointing
downslope.

Lava also forms massive pillars that stand like Greek columns on the ocean
floor up to 45 feet tall. How these strange spires formed remains a mystery.

The best explanation suggests that the pillars were created by the slow advances of lava oozing from volcanic ridges. Several blobs of lava nestle together in a ring, leaving an empty, water–filled space in the center. The sides of these adjoining blobs form the pillar walls as the outer layers cool on contact with seawater. The insides of the blobs remain fluid until the lava flows back into the vent. The fragile blobs then collapse, somewhat like large empty eggshells, leaving hollow pillars formed from the interior walls of the ring.

Among the strangest discoveries at hot vents on the deep ocean floor were giant towers of rock called chimneys and smokers that discharge very hot water, often gray or jet black (Fig. 200). The towers built up as suspended minerals in the superhot fluid were precipitated by the icy seawater. This caused metal sulfides to build up and created towers often exceeding 30 feet in height. They apparently grow fairly rapidly. During a dive on the East Pacific Rise in December 1993, the submersible *Alvin* accidentally toppled a 33–foot–tall smoker. When the sub returned three months later, the tower had already grown back 20 feet. The largest known black smoker is a 160–foot–tall structure on the Juan de Fuca ridge off the coast of Oregon appropriately named Godzilla after the giant ape of Japanese science fiction film fame. Nearby vents gush water as hot as 750 degrees Celsius, which is kept from boiling by the crushing pressure of the abyss. The vents host a variety of species and mineral deposits (Fig. 201).

In rapidly spreading rift systems such as the East Pacific Rise south of Baja California, hydrothermal vents build prodigious forests of exotic chimneys. They spew out large quantities of hot water blackened by sulfur compounds and are appropriately named black smokers. Other vents, called white smokers, eject hot water that is milky white. Seawater seeping through the ocean crust acquires heat near magma chambers below the rifts and expels with considerable force through vents like undersea geysers. The term geyser originates from the Icelandic word *geysir,* meaning "gusher." This adequately describes a geyser's behavior because of its intermittent and explosive nature, with hot water ejected with great force.

The hydrothermal water is up to 400 degrees Celsius or more but does not boil because at these great depths the pressure is 200 to 400 atmospheres. The superhot water is rich in dissolved minerals such as iron, copper, and zinc that precipitate out upon contact with the cold water of the abyss. The sulfide minerals ejected from hydrothermal vents build tall chimney structures, some with branching pipes. The black sulfide minerals drift along in the ocean currents somewhat like thick smoke from factory smokestacks.

The openings of the vents typically range from less than $1/2$ inch to more than 6 feet across. They are common throughout the world's oceans along the midocean spreading ridge system and are believed to be the main route through which Earth's interior loses heat. The vents exhibit a strange

Figure 200 *A black smoker on the East Pacific Rise.*
(Photo by R. D. Ballard, courtesy WHOI)

phenomenon by glowing in the pitch-black dark, possibly caused by the sudden cooling of the 350-degree water, which produces a phenomenon called crystalloluminescence. As dissolved minerals crystallize and drop out of solution, they emit a weak light that is just bright enough to support photosynthesis on the very bottom of the deep sea.

About 750 miles southwest of the Galapagos Islands, along the undersea mountain chain that comprises the East Pacific Rise (Fig. 202), lies an immense lava field that recently erupted. The eruption appears to have started near the ridge crest and flowed downslope over cliffs and valleys for more than 12 miles. The volume of erupted material was nearly 4 cubic miles spread over an area of some 50,000 acres, about half the annual production of new basalt on the seafloor worldwide. This is enough lava to pave the entire U.S. interstate highway system to a depth of 30 feet. Although not the greatest eruption in geologic history, this could well be the largest basalt flow in historic times. Associated with these huge bursts of basalt are megaplumes of warm, mineral-laden water measuring up to 10 miles or more across and thousands of feet deep.

The submersible *Alvin* (Fig. 203), launched from the oceanographic research *Atlantis II,* is the workhorse for exploring the deep ocean floor. In April 1991, oceanographers aboard *Alvin* witnessed an actual eruption or its immediate aftermath on the East Pacific Rise about 600 miles southwest of Acapulco, Mexico. The scientists realized the seafloor had recently erupted because the scenery did not match photographs taken at the location 15 months earlier.

The scene showed recent lava eruptions that sizzled a community of tube worms and other animals living on the deep ocean floor 1.5 miles below the sea. Suspended particles turned seawater near the seafloor extremely murky. Prodigious streams of superhot water poured from the volcanic rocks, where lava flows scorched tube worms that had not yet decayed. A few partially covered colonies still clung to life, while hordes of crabs fed on the carcasses of dead animals.

A huge undersea eruption on the Juan de Fuca Ridge about 250 miles off the Oregon coast poured out batches of lava, creating new oceanic crust in a single convulsion. The ridge forms a border between the huge Pacific plate to the west and the smaller Juan de Fuca plate to the east (Fig. 204). Eruptions along the ridge occur when the two plates separate, allowing molten rock from the mantle to rise to the surface and form new crust. Over

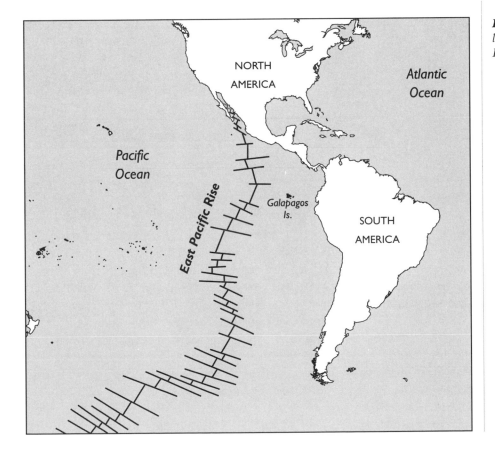

Figure 202 *The location of the East Pacific Rise.*

Figure 203 *An artist's rendition of the deep submersible* Alvin.

(Photo courtesy U.S. Navy)

time, the process of seafloor spreading carries older oceanic crust away from the ridge.

The young volcanic rocks include pillow lavas and shiny, bare basalt lacking any sediment cover. Water warmed to 50 degrees Celsius seeps out of cracks in the freshly harden basalt. In some places, tube worms have already established residency around thermal vents. The eruption appears to be related to two megaplumes discovered in the late 1980s. A string of new basaltic mounds more than 10 miles long erupted on a fracture running between the sites of the two megaplumes. The hot hydrothermal fluids along with fresh basalt gush out of the ocean floor when the ridge system cracks open and churns out more new crust.

A field of seafloor geysers off the coast of Washington State expels into the near-freezing ocean hot brine at temperatures approaching 400 degrees Celsius. Massive undersea volcanic eruptions from fissures on the ocean floor at spreading centers along the East Pacific Rise create large megaplumes of hot water. The megaplumes are produced by short periods of intense volcanic activity and can measure up to 50 to 60 miles wide.

The ridge splits open and spills out hot water while lava erupts in an act of catastrophic seafloor spreading. In a matter of a few hours, or at most a few

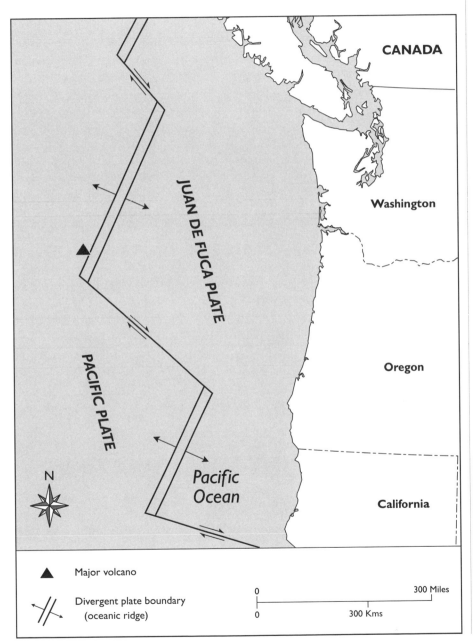

Figure 204 The location of the volcanic site on the Juan de Fuca ridge.

days, up to 100 million tons of superheated water gushes from a large crack in the ocean crust up to several miles long. When the seafloor ruptures, vast quantities of hot water held under great pressure beneath the surface violently rush out, creating colossal plumes of hot water. The release of massive amounts of superheated water beneath the sea might explain why the ocean remains salty.

Beneath the Pacific Ocean near French Polynesia, strange single-frequency notes were found emanating from clouds of bubbles billowing out of undersea volcanoes. The notes were among the purest sounds in the world, far better than those played by any musical instrument. The low frequency of the sound suggested the source had to be quite large. Further search of the ocean depths uncovered a huge swarm of bubbles. When undersea volcanoes gush out magma and scalding water, the surrounding water boils away into bubbles of steam. As the closely packed bubbles rose toward the surface, they rapidly changed shape, producing extraordinary single-frequency sound waves.

SUBMARINE SLIDES

The deep sea is not nearly as quiet as it seems. The constant tumbling of seafloor sediments down steep banks churns the ocean bottom into a murky mire. The largest slides occur on the ocean floor. As many as 40 giant submarine slides have been located around United States territory, especially near Hawaii. Submarine slides moving down steep continental slopes have buried undersea telephone cables under a thick layer of rubble. Sediments eroding out from beneath the cable leave it dangling between uneroded areas of the seabed, causing the cable to fail. A modern slide that broke a submarine cable near Grand Banks, south of Newfoundland, moved downslope at a speed of about 50 miles per hour—comparable to large terrestrial slides that devastate the landscape.

Slopes are the most common and among the most unstable landforms both on the continents as well as on the ocean floor. Under favorable conditions, the ground can give way even on the gentlest slopes, contributing to the sculpture of the landscape and seascape. Slopes are therefore inherently unstable and only temporary features over geologic time. The weakening of sediment layers due to earthquakes can cause massive subsidence. Submarine slides can be just as impressive as those on land and are responsible for much of the oceanic terrain along the outer margins of the continents.

Flow failures are among the most catastrophic types of ground failure. They consist of liquefied soil or blocks of intact material riding on a layer of liquefied sediment. Flow failures usually move dozens of feet. However, under certain geographic conditions, they can travel several miles at speeds of many miles per hour. They commonly form in loose saturated sands or silts on slopes greater than 6 percent and originate both on land and on the seafloor.

Undersea flow failures also generate large tsunamis that overrun parts of the coast. In 1929, an earthquake on the coast of Newfoundland set off a large undersea slide, triggering a tsunami that killed 27 people. On July 3, 1992, apparently a large submarine slide sent a 25-mile-long, 18-foot-high wave crashing down on Daytona Beach, Florida, overturning automobiles and injuring 75 people.

A train of three giant waves 50 feet high swept away 2,200 residents of Papua New Guinea on July 17, 1998. The disaster was originally blamed on a nearby undersea earthquake of 7.1 magnitude. However, this temblor was too small to heave up waves to such heights. Evidence collected during marine surveys of the coast implicated a submarine slide or slump of underwater sediment large enough to spawn the waves. The continental slope bears a thick carpet of sediments, which in places has slid downhill in rapid slides and slower moving slumps. The evidence on the ocean floor suggests that large tsunamis can be generated by moderate earthquakes when accompanied by submarine slides. This phenomenon makes the hazard much more dangerous than was once thought.

Submarine slides carve out deep canyons in continental slopes. They consist of dense, sediment-laden waters that move sediments swiftly along the ocean floor. These muddy waters, called turbidity currents, travel down continental slopes and transport immensely large blocks. Turbidity currents are also initiated by river discharge, coastal storms, or other currents. They deposit huge amounts of sediment that build up the continental slopes and the flat-lying abyss below.

The continental slopes plunge thousands of feet to the ocean floor and are inclined at steep angles of 60 to 70 degrees. Sediments reaching the edge of the continental shelf slide off the continental slope by the pull of gravity. Huge masses of sediment cascade down the continental slope by gravity slides that gouge out steep submarine canyons and deposit great heaps of sediment. They are often as catastrophic as terrestrial landslides and move massive quantities of sediment downslope in a matter of hours.

Submerged deposits near the base of the main island of Hawaii rank among the largest landslides on Earth. On Kilauea's south flank on the southeast coast of Hawaii, about 1,200 cubic miles of rock are moving toward the sea at speeds of up to 10 inches per year. The earth movement is presently the largest on the planet. It could ultimately lead to catastrophic sliding comparable to those of the past that have left massive piles of rubble on the ocean floor. Slides play an important role in building up the continental slope and the deep abyssal plains, making the seafloor one of the most geologically active places on Earth.

On the ocean floor lies evidence that great chunks of the Hawaiian Islands had once slid into the sea. By far, the largest example of an undersea rockslide is along the flank of a Hawaiian volcano. The slide measured roughly 1,000 cubic miles in volume and spread some 125 miles from its point of origin. The collapse of the island of Oahu sent debris 150 miles across the deep-ocean floor, churning the sea into gargantuan waves. When part of Mauna Loa Volcano collapsed and fell into the sea about 100,000 years ago, it created a tsunami 1,200 feet high that was not only catastrophic to Hawaii but might even have caused damage along the Pacific coast of North America.

The bottom of the rift valley of the Mid-Atlantic Ridge holds the remnants of a vast undersea slide at a depth of 10,000 feet that surpasses any landslide in recorded history. A large scar on one side of the submarine volcanic range indicates the mountainside gave way and slid downhill at a tremendous speed, running up and over a smaller ridge farther downslope in a manner of minutes. The slide carried nearly 5 cubic miles of rock debris to the bottom of the ocean. By comparison, this was six times more than the 1980 Mount St. Helens landslide, the largest in modern history (Fig. 205). The slide appears to have occurred about 450,000 years ago, possibly creating a gigantic tsunami 2,000 feet high.

At the Romanche Fracture Zone, intense, localized mixing is driven by submarine "waterfalls." There deep, cold water spills through a narrow cap in the Mid-Atlantic Ridge, mixing with warmer water as it goes. The phenomenon might help explain how cold, salty water mixes with warm, fresher water to form the relatively homogeneous seawater of the lower latitudes. Another process involves open ocean tides that drive water across the ridges and canyons of the ridge flank, which sets the water column undulating in waves similar to those on the surface of the ocean. When these undersea waves break, they produce an "internal surf" that drives the mixing of deeper and shallower waters.

SEA CAVES

Caves are pounded into existence by ocean waves, plowed open by flowing ice, or arise out of lava flows. They are the most spectacular examples of the dissolving power of groundwater. Over time, acidic water flowing underground dissolves large quantities of limestone, forming a system of large rooms and tunnels. Caves develop from underground channels that carry out water that seeps in from the water table. This creates an underground stream similar to how streams flow on the surface from a breached water table. The limestone landforms resulting from this process are called karst terrains, named for a region in Slovenia famous for its caves.

Figure 205 *Devastation from the 1980 eruption of Mount St. Helens, showing extensive ice and rock debris in the foreground.*

(Photo courtesy NASA)

Water gushing from an underglacier eruption carves out an enormous ice cave. Geothermal heat beneath the ice creates a large reservoir of meltwater as much as 1,000 feet deep. A ridge of rock acts as a dam to hold back the water. The sudden breakage of the dam causes the flow of water to form a long channel under the ice. Outwash streams of meltwater flowing from a glacier also carve ice caves that can be followed far upstream.

Figure 206 *The entrance to Thurston lava tube in First Twin Crater, Halemaumau Volcano, Hawaii.*

(Photo by H. T. Sterns, courtesy USGS)

If a stream of lava hardens on the surface and the underlying magma continues to flow away, a long tunnel, called a lava tube or cave, is formed (Fig. 206). These caves can reach several tens of feet across and extend for hundreds of feet. In exceptional cases, they might extend for up to 12 miles in length. The caves might be partially or completely filled with pyroclastic materials or sediments that washed in through small fissures. Sometimes the walls and roof of lava caves are adorned with stalactites, and the floor is covered with stalagmites composed of deposits of lava.

Caves also develop in sea cliffs (Fig. 207) by the ceaseless pounding of the surf or by groundwater flow through an undersea limestone formation hollowed out as the water empties into the ocean. Wave action on limestone promontories with zones of differential hardness create sea arches, such as Needle's Eye on Gibraltar Island in western Lake Erie. A major storm at sea erodes the tall cliffs landward several tens of feet. Sometimes the pounding of the surf punches a hole in the chalk to form a sea arch.

Sinkholes (Fig. 208) form when the upper surface of a limestone formation collapses, forming a deep depression on the surface. Blue holes are submerged sinkholes in the sea that appear dark blue because of their great depth. Many blue holes dot the shallow waters surrounding the Bahama Islands southwest of Florida. They formed during the ice ages when the ocean dropped several hundred feet, exposing the ocean floor well above sea level. The sea lowered in response to the growing ice sheets

Figure 207   *Sea cave cut into siltstone, Chinitna district, Cook Inlet region, Alaska.*

(Photo by A. Grantz, courtesy USGS)

Figure 208   *Possibly the
nation's largest sinkhole,
which measures 425 feet
long, 350 feet wide, and
150 feet deep, is in
Shelby County, central
Alabama.*

(Photo courtesy USGS)

that covered the northern regions of the world, locking up huge quantities of the world's water.

During its exposure on dry land, acidic rainwater seeping into the seabed dissolved the limestone bedrock, creating immense subterranean caverns. Under the weight of the overlying rocks, the roofs of the caverns collapsed, forming huge gaping pits. At the end of the last ice age, when the ice sheets melted and the seas returned, they inundated the area and submerged the sinkholes. Blue holes can be very treacherous because they often have strong eddy currents or whirlpools during incoming or outgoing tides that can capsize an unwary boat.

On Mexico's Yucatán Peninsula is a bizarre realm of giant caverns linked by miles of twisting passages 100 feet below the sea. The underlying limestone is honeycombed with long tunnels, some several miles in length, and huge caverns that could easily hold several houses. The karst terrain gives birth to underwater caves and sinkholes. The sinkholes formed when the upper surface of a limestone formation collapsed, exposing the watery world below. The sinkholes provide access to a vast subterranean world well below the surface of Earth.

As with surface caves, the Yucatán caverns contain a profusion of icicle-shaped formations of stalactites hanging from the ceiling and stalagmites rising from the floor. The formations also include delicate, hollow stalactites called soda straws that took millions of years to create. Fish, crustaceans, and other small, primitive creatures live in the darkest recesses of the caves. Many are blind as a result of generations living without light, thereby making their eyes useless appendages. These caves represent almost an entirely new ecosystem, filled with some of the most unusual life-forms found on Earth.

In the deep dark passages of Movile Cave 60 feet below ground in southern Romania are strange, previously unknown creatures, including spiders, beetles, leeches, scorpions, and centipedes. The cave is a closed subterranean ecosystem sealed off from the surface and nourished by hydrogen sulfide rising from Earth's interior. Bacteria at the bottom of the food chain metabolize hydrogen sulfide in a process called chemosynthesis.

The bizarre animals that occupy the cave evolved over the past 5 million years. They live with little oxygen and absolutely no light. As a result, they lack pigmentation and eyesight. The cave, which winds beneath 150 square miles of countryside, began when the Black Sea dropped precipitously some 5.5 million years ago. The cave developed in a limestone formation when the waters began rising again. It was sealed off from the outside world when clay impregnated the limestone, making it watertight, and when thick layers of wind-driven sediment were deposited on top during the ice ages.

SEAFLOOR CRATERS

Because water covers more than 70 percent of Earth's surface, most meteorites land in the ocean. Several sites on the seafloor have been identified as possible marine impact craters. An asteroid or comet landing in the ocean would produce a conical-shaped curtain of water as billions of tons of seawater splash high into the air. The atmosphere would become oversaturated with water vapor. Thick cloud banks would shroud the planet, blocking out the sun. Massive tsunamis would race outward from the impact site and traverse completely around the world. When striking seashores, they would travel hundreds of miles inland, devastating everything in their paths.

About 65 million years ago, a large meteorite supposedly struck Earth, creating a crater at least 100 miles wide. The debris sent the planet into environmental chaos. This catastrophe might have caused the demise of the dinosaurs along with 70 percent of all other species. Since no crater has been found on the continents, the meteorite probably landed in the ocean. If so, millions of years of sedimentation would have erased all signs of it.

Much of the search for the dinosaur killer impact site has been concentrated around the Caribbean area (Fig. 209). There thick deposits of wave-deposited rubble exist along with melted and crushed rock ejected from the crater. The most suitable site for the proposed crater is the Chicxulub impact structure, one of the largest known on Earth. It measures from 110 to 185 miles wide. It is named for a small village at its center that means "the devil's tail" in Mayan. The crater lies beneath 600 feet of sedimentary rock on the northern coast of the Yucatán Peninsula.

If the meteorite landed on the seabed just offshore, 65 million years of sedimentation would have long since buried it under thick deposits of sand and mud. Furthermore, a splashdown in the ocean would have created an enormous tsunami that would scour the seafloor and deposit its rubble on nearby shores. The impact would have set off tremendous earthquakes, whose shaking would have sent sediment from shallow waters sliding off the continent shelf. When the ooze settled on the deep ocean floor, it could have blanketed a region as large as 1.5 million square miles, an area more than twice the size of Alaska.

Figure 209 Possible impact structures in the Caribbean area that might have ended the Cretaceous period.

The impact structure dates precisely to the end of the dinosaur era about 65 million years ago. The impact is thought to have caused the giant beasts'

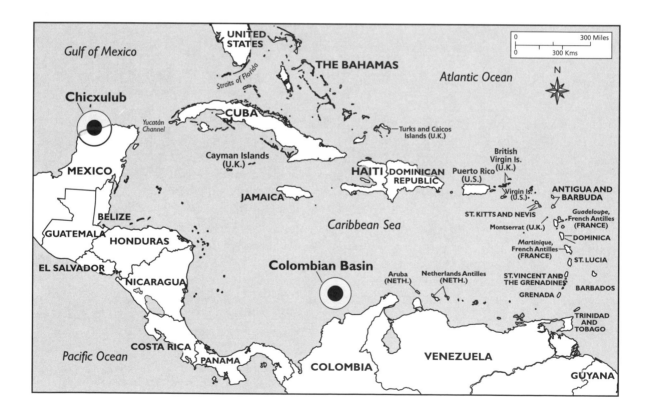

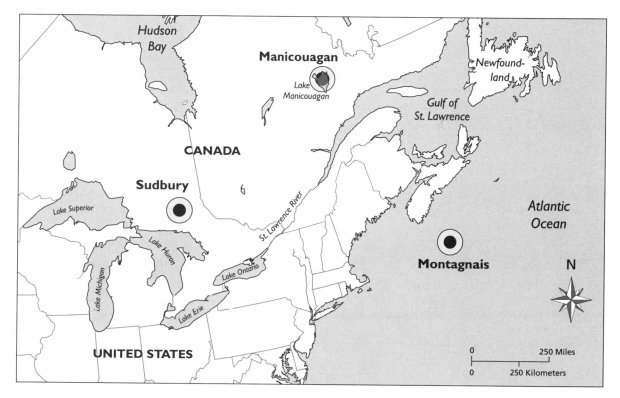

extinction. The buried crater is outlined by an unusual concentration of sink-holes. The impact structure forms a circular fracture system that acts as an underground river. The cavity formation in the sinkholes extends to a depth of about 1,000 feet. Its permeability causes the ring to act as a conduit carrying groundwater to the sea.

The most pronounced undersea impact crater known is the 35-mile-wide Montagnais structure (Fig. 210) 125 miles off the southeast coast of Nova Scotia. Oil companies exploring for petroleum in the area discovered the circular formation. The crater is 50 million years old. It closely resembles craters on dry land, only its rim is 375 feet beneath the sea and the crater floor is 9,000 feet deep. A large meteorite up to 2 miles wide excavated the crater. The impact raised a central peak similar to those seen inside craters on the Moon.

The impact structure also contained rocks melted by a sudden shock. Such an impact would have sent a tremendous tsunami crashing down onto nearby shores. Because of its size and location, the crater was thought to be a likely candidate for the source of the North American tektites (Fig. 211)

Figure 210   The location of the Montagnais crater off Nova Scotia, Canada.

strewn across the American West. Unfortunately, it proved to be several million years too young to have created these tektites. However, the ocean is vast, and better candidates might some day reveal themselves.

A meteorite slamming into the Atlantic Ocean along the Virginia coast about 40 million years ago released a huge wave that pounded the adjacent shoreline. Apparently, the tsunami gouged out of the seafloor a 5,000-square-mile region about the size of Connecticut. When the meteorite crashed into the submerged continental shelf, it created a wave that ripped the seafloor into an enormous number of large boulders. A layer of 3-foot boulders 200 feet thick laid in three locations, buried under 1,200 feet of sediment. Within the boulder layer were mineral grains showing shock features and glassy rocks called tektites formed when a meteorite blasted the seafloor and flung the molten rock in all directions.

A large meteorite impact might have created the Everglades on the southern tip of Florida. The Everglades is a swamp and forested area surrounded by an oval-shaped system of ridges upon which rest most of southern Florida's cities. A giant coral reef, dating about 6 million years old, lies beneath the rim surrounding the Everglades. The coral reef probably formed

Figure 211   *A North American tektite found in Texas in November 1985, showing surface erosional and corrosional features.*

(Photo by E. C. T. Chao, courtesy USGS)

around the circular basin gouged out by the meteorite impact. A thick layer of limestone surrounding the area and laid down about 40 million years ago is suspiciously missing over most of the southern part of the Everglades. Apparently, a large meteorite slammed into limestones submerged under 600 feet of water and fractured the rocks. The impact would also have generated an enormous tsunami that swept the debris far out to sea.

About 2.3 million years ago, a major asteroid appears to have impacted the ocean floor in the Pacific Ocean roughly 700 miles westward of the tip of South America. Although no crater was found, an excess of iridium (a rare isotope of platinum found in abundance on meteorites) in sand-sized bits of glassy rock existed in the area, suggesting an extraterrestrial origin. The impact created at least 300 million tons of debris, consistent with an object about a half mile in diameter. The blast from the impact would have been equal to that of all the nuclear arsenals in the world, with devastating consequences for the local ecology. Moreover, geologic evidence suggests that Earth's climate changed dramatically between 2.2 and 2.5 million years ago, when glaciers covered large parts the Northern Hemisphere.

Lying in the middle of the desert sands of Australia are curious looking boulders of exotic rock called drop stones that measure as much as 10 feet across and originated from great distances away. These large, out-of-place boulders strewn across Australia's central desert far from their sources suggest they rafted out to sea on slabs of drift ice when a large inland waterway invaded the continent during the Cretaceous period.

When the icebergs melted, the huge rocks simply dropped to the ocean floor, where their impacts produced craterlike depressions in the soft sediment layers. Evidence of ice-rafted boulders also exists in glacial soils in other parts of the world, including the Canadian Arctic and Siberia. Similar boulders were found in sediments from other warm periods as well. Even today, the same ice-rafting process continues in the Hudson Bay area.

UNDERSEA EXPLOSIONS

The most explosive volcanic eruption in recorded history occurred during the 17th century B.C. on the island of Thera 75 miles north of Crete in the Mediterranean Sea. The magma chamber beneath the island apparently flooded with seawater. Like a gigantic pressure cooker, the volcano blew its lid. The volcanic island collapsed into the emptied magma chamber, forming a deep water-filled caldera that covered an area of 30 square miles. The collapse of Thera also created an immense tsunami that battered the shores of the eastern Mediterranean.

Krakatoa lies in the Sundra Strait between Java and Sumatra, Indonesia. On August 27, 1883, a series of four powerful explosions ripped the island apart. The explosions were probably powered by the rapid expansion of steam, generated when seawater entered a breach in the magma chamber. Following the last convulsion, most of the island caved into the emptied magma chamber. This created a large undersea caldera more than 1,000 feet below sea level, resembling a broken bowl of water with jagged edges protruding above the surface of the sea.

The first hydrogen bomb test was conducted on November 1, 1952, on Elugelab atoll, in the Eniwetok Lagoon in the South Pacific. The nuclear device named Mike measured 22 feet long and 5 feet wide and weighed about 65 tons. It had an explosive force estimated at 10 megatons of TNT. When Mike was detonated, the fireball expanded to more than 3 miles in diameter in less than a second (Fig. 212). Millions of gallons of seawater instantly boiled into steam. After the clouds cleared, Elugelab was no more. A huge crater was blown into the ocean floor 1 mile wide and 1,500 feet deep.

Another type of crater on the bottom of the ocean was formed by a natural seafloor explosion. In 1906, sailors in the Gulf of Mexico witnessed a massive gas blowout that sent mounds of bubbles to the surface. The area is known for its reservoirs of hydrocarbons that might have caused the explosion. Pockets of gases lie trapped under high pressure deep beneath the floor of the ocean. As the pressure increases, the gases explode undersea, spreading debris in all directions and producing huge craters on the ocean floor. The gases rush to the surface in great masses of bubbles that burst in the open air, resulting in a thick foamy froth on the surface of the ocean.

Further exploration of the site yielded a large crater on the ocean floor, lying in 7,000 feet of water southeast of the Mississippi River delta. The elliptical hole measured 1,300 feet long, 900 feet wide, and 200 feet deep and sat atop a small hill. Downslope laid more than 2 million cubic yards of ejected sediment. Apparently, gases seeped upward along cracks in the seafloor and collected under an impermeable barrier. Eventually, the pressure forced the gas to blow off its cover, forming a huge blowout crater.

In the Gulf of Mexico, as well as in other parts of the world, the seabed overlies thick salt deposits formed when the sea evaporated during a warmer climate. A sedimentary dome is created when the crust is heaved upward often due to salt tectonics. Since salt buried in the crust from ancient seabeds is lighter than the surrounding rocks, it slowly rises toward the surface, bulging the overlying strata upward. Often oil and gas is trapped in these structures, and petroleum geologists spend much time looking for salt domes.

Figure 212 *The first hydrogen bomb test, named Mike, on Elugelab Atoll on November 1, 1955.*

(Photo courtesy Defense Nuclear Agency)

The bottom of the Gulf is lined with a layer of anhydrite, an anhydrous (water-saturated) calcium sulfate common in evaporite deposits. The anhydrite forms an impervious stratum to the buildup of gas beneath the surface. When the building gas pressure overcomes the barrier, gases rush toward the surface, forming a froth on the open ocean. A ship sailing into such a foamy sea would lose all buoyancy because it is no longer supported by seawater. As

Figure 213 *The Bermuda Triangle has been blamed for mysterious disappearances of ships and planes.*

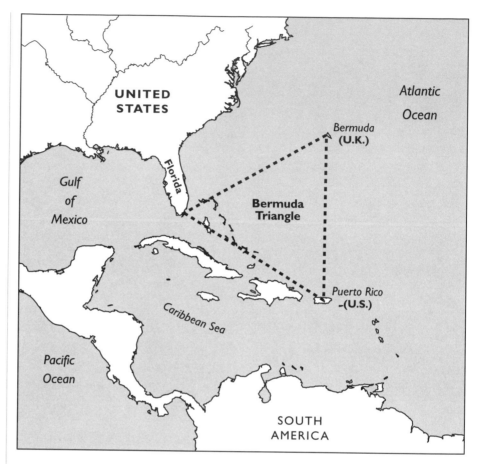

a result the ship would immediately sink to the bottom. An airplane flying overhead might stall out, choking on the pall of poisonous gases. Perhaps these phenomena might explain the strange disappearance of ships and aircraft in the Caribbean around a region known as the Bermuda Triangle (Fig. 213), one of many unsolved mysteries of the sea.

CONCLUSION

Marine geology, also called geological oceanography, is an important field of geology that explores the ocean floor and many of its features. It includes the study of Earth at the sea's edge and deep below its surface. More recent discoveries on the ocean floor have led to the development of the theory of plate tectonics. The jostling of great crustal plates is responsible for much of the dynamic geology of the seabed.

The importance of plate tectonics in marine geology cannot be overstated because a basic knowledge is necessary for understanding the geologic activity of three-quarters of Earth's surface. Much of the seascape was sculpted by the interactions of huge slabs of crust. The great undersea mountain ranges arose from the action of seafloor spreading. The deep-sea trenches associated with island arcs resulted from plate subduction. Therefore, without plate tectonics, the seabed would simply be a barren desert.

GLOSSARY

aa lava (AH-ah)   Hawaiian name for blocky basalt lava

abrasion   erosion by friction, generally caused by rock particles carried by running water, ice, and wind

absorption   the process by which radiant energy incident on any substance is retained and converted into heat or other forms of energy

abyss (ah-BIS)   the deep ocean, generally over a mile in depth

accretion   the accumulation of celestial dust by gravitational attraction into a planetesimal, asteroid, moon, or planet

advection currents   the horizontal movement of air or water

aerosol   a mass of minute solid or liquid particles dispersed in the air

albedo   the amount of sunlight reflected from an object and dependent on color and texture

alluvium (ah-LUE-vee-um)   stream-deposited sediment

alpine glacier   a mountain glacier or a glacier in a mountain valley

ammonite (AM-on-ite)   a Mesozoic cephalopod with flat, spiral shells

andesite (AN-di-zite)   an intermediate type of volcanic rock between basalt and rhyolite

annelid (A-nil-ed)   wormlike invertebrate characterized by a segmented body with a distinct head and appendages

aquifer (AH-kwe-fer)   a subterranean bed of sediments through which groundwater flows

Archean (ar-KEY-an)   major eon of the Precambrian from 4.0 to 2.5 billion years ago

arthropod (AR-threh-pod)   the largest group of invertebrates, including crustaceans and insects, characterized by segmented bodies, jointed appendages, and exoskeletons

ash fall   the fallout of small, solid particles from a volcanic eruption cloud

asteroid   a rocky or metallic body, orbiting the Sun between Mars and Jupiter, and left over from the formation of the solar system

asthenosphere (as-THE-nah-sfir)   a layer of the upper mantle from about 60 to 200 miles below the surface that is more plastic than the rock above and below and might be in convective motion

atmospheric   the weight per unit area of the total mass pressure air above a given point; also called barometric pressure

back–arc basin   a seafloor spreading system of volcanoes caused extension behind an island arc that is above a subduction zone

Baltica (BAL-tik-ah)   an ancient Paleozoic continent of Europe

barrier island   a low, elongated coastal island that parallels the shoreline and protects the beach from storms

basalt (bah-SALT)   a dark, volcanic rock rich in iron and magnesium and usually quite fluid in the molten state

benthic front   ocean currents that flow along the bottom of the deep ocean

biogenic   sediments composed of the remains of plant and animal life such as shells

biomass   the total mass of living organisms within a specific habitat

biosphere   the living portion of the Earth that interacts with all other biological and geologic processes

black smoker   superheated hydrothermal water rising to the surface at a midocean ridge; the water is supersaturated with metals; when exiting through the seafloor, the water quickly cools and the dissolved metals precipitate, resulting in black, smokelike effluent

blueschist (BLUE-shist)   metamorphosed rocks of subducted ocean crust exposed on land

brachiopod (BRAY-key-eh-pod)   marine, shallow-water invertebrate with bivalve shells similar to mollusks and plentiful in the Paleozoic

bryozoan (bry-eh-ZOE-an)   a marine invertebrate that grows in colonies and characterized by a branching or fanlike structure

calcite   a mineral composed of calcium carbonate

caldera (kal-DER-eh)   a large, pitlike depression at the summits of some volcanoes and formed by great explosive activity and collapse

calving   formation of icebergs by breaking off of glaciers entering the ocean

carbonaceous (KAR-beh-NAY-shes)   a substance containing carbon, namely sedimentary rocks such as limestone and certain types of meteorites

carbonate   a mineral containing calcium carbonate such as limestone and dolostone

carbon cycle   the flow of carbon into the atmosphere and ocean, the conversion to carbonate rock, and the return by volcanoes

Cenozoic (sin-eh-ZOE-ik)   an era of geologic time comprising the last 65 million years

cephalopod (SE-feh-lah-pod)   marine mollusks including squids, cuttlefish, and octopuses that travel by expelling jets of water

chalk   a soft form of limestone composed chiefly of calcite shells of microorganisms

chemosynthesis   manufacturing of organic compounds by energy from chemical reactions, such as those on the deep-sea floor near hydrothermal vents

chert   an extremely hard, cryptocrystalline quartz rock resembling flint

chondrule (KON-drule)   rounded granules of olivine and pyroxine found in stony meteorites called chondrites

circum-Pacific belt   active seismic regions around the rim of the belt Pacific plate coinciding with the Ring of Fire

circumpolar   traveling around the world from pole to pole

climate   the average course of the weather for a certain region over time

coastal storm   a cyclonic, low-pressure system moving along a coastal plain or just offshore; it causes north to northeast winds over the land, and along the Atlantic seaboard it is called a northeaster

coelacanth (SEE-leh-kanth)   a lobe-finned fish originating in the Paleozoic and presently living in deep seas

coelenterate (si-LEN-the-rate)   a multicellular marine organism, such as a jellyfish or coral

condensation   the process whereby a substance changes from the vapor phase to the liquid or solid phase; the opposite of evaporation

conductivity   the property of transmitting a quality

continent   a landmass composed of light, granitic rock that rides on the denser rocks of the upper mantle

continental drift   the concept that the continents drift across the surface of Earth

continental glacier   an ice sheet covering a portion of a continent

continental margin   the area between the shoreline and the abyss that represents the true edge of a continent

continental shelf   the offshore area of a continent in shallow sea

continental shield   ancient crustal rocks upon which the continents grew

continental slope   the transition from the continental shelf to the deep-sea basin

convection   a circular, vertical flow of a fluid medium by heating from below; as materials are heated, they become less dense and rise, then they cool down, and become more dense and sink

convergent plate margin   the boundary between crustal plates where the plates come together; generally corresponds to the deep-sea trenches where old crust is destroyed in subduction zones

coral   a large group of shallow-water, bottom-dwelling marine invertebrates comprising reef-building colonies common in warm waters

cordillera (kor-dil-ER-ah)   a range of mountains that includes the Rockies, Cascades, and Sierra Nevada in North America and the Andes in South America

core   the central part of Earth, consisting of a heavy iron-nickel alloy; also a cylindrical rock sample drilled through the crust

Coriolis effect   the apparent force that deflects the wind or a moving object, causing it to curve in relation to the rotating Earth

craton (CRAY-ton)   the ancient, stable interior of a continent, usually composed of the precambrian rocks

crinoid (KRY-noid)   an echinoderm with a flowerlike body atop a long stalk of calcite disks

crosscutting   a body of rocks cutting across older rock units

crust   the outer layers of a planet's or a moon's rocks

crustacean (kres TAY-shen)   an arthropod characterized by two pairs of antenna-like appendages forward of the mouth and three pairs behind it, including shrimps, crabs, and lobsters

crustal plate   a segment of the lithosphere involved in the interaction of other plates in tectonic activity

delta   the wedge-shaped pile of sediments deposited at the mouth of a river

density   the mass per unit volume

desiccated basin (DEH-si-kay-ted)   a basin formed when an ancient sea evaporated

diapir (DIE-ah-per)   the buoyant rise of a molten rock through heavier rock

diatom (DIE-ah-tom)   microplants whose fossil shells form siliceous sediments called diatomaceous earth

diatomite   an ultra-fine-grained siliceous earth composed mainly of diatom cell walls

divergent plate margin   the boundary between lithospheric plates where they separate; it generally corresponds to midocean ridges where new crust is formed by the solidification of liquid rock rising from below

dynamo effect   the creation of Earth's magnetic field by rotational, thermal, chemical, and electrical differences between the solid inner core and the liquid outer core

earthquake   the sudden rupture of rocks along active faults in response to geologic forces within Earth

East Pacific Rise   a midocean ridge spreading center running north-south along the eastern side of the Pacific; the predominant location where hot springs and black smokers were discovered

echinoderm (i-KY-neh-derm)   marine invertebrates, including starfish, sea urchins, and sea cucumbers

eon   the longest unit of geologic time, roughly about a billion years or more in duration

erosion   the wearing away of surface materials by natural agents such as wind and water

evaporation   the transformation of a liquid to a gas

evaporite (ee-VA-per-ite)   the deposition of salt, anhydrite, and gypsum from evaporation in an enclosed basin of stranded seawater

evolution   the tendency of physical and biological factors to change with time

extrusive (ik-STRU-siv)   an igneous volcanic rock ejected onto Earth's surface

fluvial (FLUE-vee-al)   stream-deposited sediment

foraminifer (FOR-eh-MI-neh-fer)   a calcium carbonate-secreting organism that lives in the surface waters of the oceans; after death, its shell forms the primary constituent of limestone and sediments deposited onto the seafloor

fossil   any remains, impressions, or traces in rock of a plant or animal of a previous geologic age

fossil fuel   an energy source derived from ancient plant and animal life that includes, coal, oil, and natural gas; when ignited, these fuels release carbon that was stored in Earth's crust for millions of years

fracture zones   narrow regions consisting of ridges and valleys parallel to spreading ridges and aligned in a stair-step shape

fumarole (FUME-ah-role)   a vent through which steam or other hot gases escape from underground, such as a geyser

gastropod (GAS-tra-pod)   a large class of mollusks, including slugs and snails, characterized by a body protected by single shell that is often coiled

geochemical   pertaining to the distribution and circulation of chemical elements in Earth's soil, water, and atmosphere

geologic column   the total thickness of geologic units in a region

geostrophic flow   ocean currents that flow perpendicular to the Coriolis flow or to the right of the boundary currents in the Northern Hemisphere

geothermal   the generation of hot water or steam by hot rocks in the Earth's interior

geyser (GUY-sir)   a spring that ejects intermittent jets of steam and hot water

glacier   a thick mass of moving ice occurring where winter snowfall exceeds summer melting

Gondwana (gone-DWAN-ah) a southern supercontinent of Paleozoic time, comprising Africa, South America, India, Australia, and Antarctica; it broke up into the present continents during the Mesozoic era

graben (GRA-bin) a valley formed by a downdropped fault block

granite a coarse-grain, silica-rich rock consisting primarily of quartz and feldspars

gravimeter an instrument that measures the intensity of Earth's gravity

greenhouse effect the trapping of heat in the lower atmosphere principally by water vapor and carbon dioxide

greenstone a green, weakly metamorphic igneous rock

groundwater water derived from the atmosphere that percolates and circulates below the surface

guyot (GHEE-oh) undersea volcano that once existed above sea level and whose top was flattened by erosion; later, subsidence caused the volcano to sink below the ocean surface, preserving its flat-top appearance

hot spot a volcanic center with no relation to a plate boundary; an anomalous magma generation site in the mantle

hydrocarbon a molecule consisting of carbon chains with attached hydrogen atoms

hydrologic cycle the flow of water from the ocean to the land and back to the sea

hydrology the study of water flow over Earth

hydrosphere the water layer at the surface of Earth

hydrothermal relating to the movement of hot water through the crust; it is the circulation of cold seawater downward through the oceanic crust toward the deeper depths of the oceanic crust where it becomes hot and buoyantly rises toward the surface

Iapetus Sea (eye-AP-i-tus) a former sea that occupied a similar area as the present Atlantic Ocean prior to the assemblage of Pangaea

ice age a period of time when large areas of Earth were covered by massive glaciers

iceberg a portion of a glacier calved off upon entering the sea

ice cap a polar cover of ice and snow

igneous rocks all rocks solidified from a molten state

impact the point on the surface upon which a celestial object lands

internal wave a wave propagating at a density boundary within the ocean rather than at the surface of the water

invertebrate an animal with an external skeleton such as shellfish and insects

island arc volcanoes landward of a subduction zone, parallel to a trench of a subducting plate and above the plate's melting zone

isostasy (eye-SOS-tah-see) a geologic principle that states that Earth's crust is buoyant and rises and sinks depending on its density

landslide   a rapid downhill movement of earth materials triggered by earthquakes and severe weather

Langmuir circulation   near-surface alternating vortices aligned downwind, generated by the interaction of waves and mean shear currents

Laurasia (lure-AY-zha)   a northern supercontinent of Paleozoic time, consisting of North America, Europe, and Asia

Laurentia (lure-IN-tia)   an ancient North American continent

lava   molten magma that flows out onto the surface

limestone   a sedimentary rock composed of calcium carbonate that is secreted from seawater by invertebrates and whose skeletons compose the bulk of deposits

lithosphere (LI-tha-sfir)   the rocky outer layer of the mantle that includes the terrestrial and oceanic crusts; the lithosphere circulates between Earth's surface and mantle by convection currents

lithospheric plate   a segment of the lithosphere, the upper-layer plate of the mantle, involved in the interaction of other plates in tectonic activity

lysocline   the ocean depth below which the rate of dissolution just exceeds the rate of deposition of the dead shells of calcareous organisms

magma   a molten rock material generated within Earth and that is the constituent of igneous rocks

magnetic field reversal   a reversal of the north-south polarity of Earth's magnetic poles

magnetometer   a devise used to measure the intensity and direction of the magnetic field

manganese nodule   a cobble-shaped ore on the deep-sea floor, which is rich in manganese and iron

mantle   the part of a planet below the crust and above the core, composed of dense rocks that might be in convective flow

massive sulfides   sulfide metals deposited from hydrothermal solutions

megaplume   a large volume of mineral-rich warm water above an oceanic rift

Mesozoic (MEH-zeh-ZOE-ik)   literally the period of middle life, referring to a period between 250 and 65 million years ago

metamorphism (me-teh-MORE-fi-zem)   recrystallization of previous igneous, metamorphic, and sedimentary rocks created under extreme temperatures and pressures without melting

meteorite   a metallic or stony celestial body that enters Earth's atmosphere and impacts onto the surface

microplate   a small block of ocean crust surrounded by major plates

Mid-Atlantic Ridge   the seafloor-spreading ridge that marks the extensional edge of the North and South American plates to the west and the Eurasian and African plates to the east

midocean ridge a submarine ridge along a divergent plate boundary where a new ocean floor is created by the upwelling of mantle material

Mohorovicic discontinuity/Moho (MOE-HOE) the boundary between the crust and mantle discovered by Andrija Mohorovicic

mollusk (MAH-lusc) a large group of invertebrates, including snails, clams, squids, and extinct ammonites, characterized by an internal and external shell surrounding the body

natural selection the process by which evolution selects species for survival or extinction depending on the environment

nontransform fault small offsets with overlapping ridge tips at faults that offset the Mid-Atlantic Ridge

nuée ardente (NU-ay ARE-don) an avalanche of glowing clouds of ash and pyroclastics

olivine (AH-leh-vene) a green, iron-magnesium silicate common in intrusive and volcanic rocks

ophiolite (oh-FI-ah-lite) oceanic crust thrust upon continents by plate tectonics

ore body the accumulation of metal-bearing ores where the hot hydrothermal water moving upward toward the surface mixes with cold sea water penetrating downward

orogeny (oh-RAH-ja-nee) an episode of mountain building by tectonic activity

outgassing the loss of gas from within a planet as opposed to degassing, the loss of gas from meteorites

overthrust a thrust fault in which one segment of crust overrides another segment for a great distance

oxidation the chemical combination of oxygen with other elements

pahoehoe (pah-HOE-ay-hoe-ay) a Hawaiian term for ropy basalt lava

paleomagnetism the study of Earth's magnetic field, including the position and polarity of the poles in the past

paleontology (pay-lee-on-TAH-logy) the study of ancient life-forms, based on the fossil record of plants and animals

Paleozoic (PAY-lee-eh-ZOE-ic) the period of ancient life, between 540 and 250 million years ago

Pangaea (pan-GEE-a) a Paleozoic supercontinent that included all the lands of the Earth

Panthalassa (pan-THE-lass-ah) the global ocean that surrounded Pangaea

peridotite the most common ultramafic rock type in Earth's mantle

period a division of geologic time longer than an epoch and included in an era

photosynthesis the process by which plants form carbohydrates from carbon dioxide, water, and sunlight

phytoplankton   marine or freshwater microscopic, single-celled, freely drifting plant life

pillow lava   lava extruded on the ocean floor giving rise to tabular shapes

plate tectonics   the theory that accounts for the major features of the Earth's surface in terms of the interaction of lithospheric plates

polarity   a condition in which a substance exhibits opposite properties such as electric charges or magnetic fields

precipitation   products of condensation that fall from clouds as rain, snow, hail, or drizzle; also the deposition of rocks from seawater

primary producer   the lowest member of a food chain

radiogenic   pertaining to something produced by radioactive decay, such as heat

radiolarian   a microorganism with shells made of silica comprising a large component of siliceous sediments

radiometric dating   determining the age of an object by radiometrically and chemically analyzing its stable and unstable radioactive elements

reef   the biological community that lives at the edge of an island or continent; the shells from dead organisms form a limestone deposit

regression   a fall in sea level, exposing continental shelves to erosion

rhyolite (RYE-eh-lite)   a potassium-feldspar–rich volcanic rock equivalent to granite

ridge crest   an axis of midocean volcanoes aligned along the edge of two plates extending away from each other

rift valley   the center of an extensional spreading, where continental or oceanic plate separation occurs

Ring of Fire   a belt of subduction zones around the Pacific plate related to volcanic activity

Rodinia   a Precambrian supercontinent whose breakup sparked the Cambrian explosion of species

seafloor spreading   a theory that the ocean floor is created by the separation of lithospheric plates along midocean ridges, with new oceanic crust formed from mantle material that rises from the mantle to fill the rift

seamount   a submarine volcano that never reaches the surface of the sea

seawall   a structure built to protect against shore erosion

seaward bulge   the elevated seaward bulge produced by the bending of the subducting plate

sedimentation   the deposition of sediments

seiche   the oscillation of water in a bay

seismic (SIZE-mik)   pertaining to earthquake energy or other violent ground vibrations

seismic sea wave   an ocean wave generated by an undersea earthquake or volcano; also called tsunami

shield   areas of the exposed Precambrian nucleus of a continent

shield volcano   a broad, low-lying volcanic cone built up by lava flows of low viscosity

sonar   an instrument for measuring the ocean floor with sound waves

sounding   the measurement of water depth with weighted lines

spherules   small, spherical, glassy grains found on certain types of meteorites, on lunar soils, and at large meteorite impact sites

storm surge   an abnormal rise of the water level along a shore as a result of wind flow in a storm

stratification   a pattern of layering in sedimentary rocks, lava flows, water, or materials of different composition or density

striae (STRY-aye)   scratches on bedrock made by rocks embedded in a moving glacier

stromatolite (stro-MAT-eh-lite)   a calcareous structure built by successive layers of bacteria or algae and that has existed for the past 3.5 billion years

subduction zone   a region where an oceanic plate dives below a continental plate into the mantle; ocean trenches are the surface expression of a subduction zone

submarine canyon   a deep gorge residing underwater and formed by the underwater extensions of rivers

subsidence   the compaction of sediments due to the removal of fluids

surge glacier   a continental glacier heading toward the sea at a high rate of advance during certain times

symbiosis   the union of two dissimilar organisms for mutual benefit

tectonic activity   the formation of Earth's crust by large-scale movements throughout geologic time

tectonics (tek-TAH-niks)   the history of Earth's larger features (rock formations and plates) and the forces and movements that produce them; see plate tectonics

tephra (TE-fra)   solid material ejected into the air by a volcanic eruption

Tethys Sea (TEH-this)   the hypothetical, midlatitude region of the oceans separating the northern and southern continents of Laurasia and Gondwana several hundred million years ago

thermal   the amount of heat conducted per unit of time through any cross section of a substance, dependent on the temperature gradient at that section and the area of the section

thermocline   the boundary between cold and warm layers of the ocean

tidal friction   the loss of energy through heating caused by the movements associated with the tides

tide   a bulge in the ocean produced by the Sun's and Moon's gravitational forces on the Earth's oceans; the rotation of the Earth beneath this bulge causes the rising and lowering of the sea level

transform fault   a fracture in Earth's crust along which lateral movement occurs; a common feature of the midocean ridges created in the line of seafloor spreading

transgression   a rise in sea level that causes flooding of the shallow edges of continental margins

traps   a series of massive lava flows that resembles a staircase

trench   a depression on the ocean floor caused by plate subduction

tsunami (sue-NAH-me)   a seismic sea wave produce by an undersea or nearshore earthquake or volcanic eruption

tubeworm   a retractable wormlike animal living within a long stalk near hydrothermal vents

turbidite   a slurry of mud that periodically slides down often gentle slopes toward the deep-sea floor

typhoon   a severe tropical storm in the Western Pacific similar to a hurricane

upwelling   the upward convection of water currents

volcanism   any type of volcanic activity

volcano   a fissure or vent in the crust through which molten rock rises to the surface to form a mountain

white smoker   a hydrothermal vent on the deep-sea floor similar to a black smoker but that produces a white effluent

BIBLIOGRAPHY

THE BLUE PLANET

Allegre, Claud J. and Stephen H. Snider. "The Evolution of the Earth." *Scientific American* 271 (October 1994): 66–75.

Dalziel, Ian W. D. "Earth Before Pangaea." *Scientific American* 272 (January 1995): 58–63.

Gurnis, Michael. "Sculpting the Earth from the Inside Out." *Scientific American* 284 (March 2001): 40–47.

Hoffman, Paul F. and David P. Schrag. "Snowball Earth." *Scientific American* 282 (January 2000): 68–75.

Kerr, Richard A. "A Refuge for Life on Snowball Earth." *Science* 288 (May 26, 2000): 1316.

Knauth, Paul. "Ancient Sea Water." *Nature* 362 (March 25, 1993) 290–291.

Knoll, Andrew H. "End of the Proterozoic Eon." *Scientific American* 265 (October 1991): 64–73.

Moores, Eldridge. "The Story of Earth." *Earth* 6 (December 1996): 30–33.

Motani, Ryosuke. "Rulers of the Jurassic Seas." *Scientific American* 283 (December 2000): 52–58.

Szelc, Gary. "Where Ancient Seas Meet Ancient Sand." *Earth* 5 (December 1997): 78–81.

Vermeij, Geerat J. "The Biological History of a Seaway." *Science* 260 (June 11, 1993): 1603–1604.

Weiss, Peter. "Land Before Time." *Earth* 8 (February 1998): 29–33.

York, Derek. "The Earliest History of the Earth." *Scientific American* 268 (January 1993): 90–96

Zimmer, Carl. "Ancient Continent Opens Window on the Early Earth." *Science* 286 (December 17, 1999): 2254–256.

MARINE EXPLORATION

Broad, William J. "Life Springs Up in Ocean's Volcanic Vents, Deep Divers Find." *The New York Times* (October 19, 1993): C4.

DiChristina, Mariette. "Science at Sea." *Popular Science* 251 (August 1997): 82–83.

Hoffman, Kenneth A. "Ancient Magnetic Reversals: Clues to the Geodynamo." *Scientific American* 258 (May 1988): 76–83.

Kerr, Richard A. "Coming up Short in Crustal Quest." *Science* 254 (December 6, 1991): 1456–1457.

Monastersky, Richard. "A New View of Earth." *Science News* 148 (December 16, 1995): 410–411.

———. "Drilling Shortcut Penetrates Earth's Mantle." *Science News* 143 (February 20, 1993): 117.

Orange, Daniel L. "Mysteries of the Deep." *Earth* 5 (December 1996): 42–45.

Pratson, Lincoln F. and William F. Haxby. "Panoramas of the Seafloor." *Scientific American* 276 (June 1997): 82–87.

Weisburd, Stefi. "Sea-Surface Shape by Satellite." *Science News* 129 (January 18, 1986): 37.

Wood, Denis. "The Power of Maps." *Scientific American* 268 (May 1993): 89–93.

Yulsman, Tom. "The Seafloor Laid Bare." *Earth* 5 (June 1996): 45–51.

Zimmer, Carl. "Inconsistent Field." *Discover* 15 (February 1994): 26–27.

THE DYNAMIC SEAFLOOR

Berner, Robert A. and Antonio, C. Lasaga. "Modeling the Geochemical Carbon Cycle." *Scientific American* 260 (March 1989): 74–81.

Cann, Joe and Cherry Walker. "Breaking New Ground on the Ocean Floor." *New Scientist* 139 (October 30, 1993): 24–29.

Cathles, Flanagan, Ruth. "Sea Change." *Earth* 8 (February 1998): 42–47.

Gordon, Richard G. and Seth Stein. "Global Tectonics and Space Geodesy." *Science* 256 (April 17, 1992): 333–341.

Green, D. H., S. M. Eggins, and G. Yaxley. "The Other Carbon Cycle." *Nature* 365 (September 16, 1993): 210–211.

Howell, David G. "Terranes." *Scientific American* 252 (November 1985): 116–125.

Kerr, Richard A. "Ocean Crust Role in Making Seawater." *Science* 239 (January 15, 1988): 260.

Lawrence, M., III. "Scales and Effects of Fluid Flow in the Upper Crust." *Science* 248 (April 20, 1990): 323–328.

Monastersky, Richard. "Drilling Shortcut Penetrates Earth's Mantle." *Science News* 143 (February 20, 1993): 117.

Mutter, John C. "Seismic Images of Plate Boundaries." *Scientific American* 254 (February 1986): 66–75.

Stone, Richard. "Black Sea Flood Theory to Be Tested." *Science* 283 (February 12, 1999): 915–916.

RIDGES AND TRENCHES

Appenzeller, Tim. "How Vanished Oceans Drop an Anchor." *Science* 270 (November 17, 1995): 1122.

Bonatti, Enrico and Kathleen Crane. "Oceanic Fracture Zones." *Scientific American* 250 (May 1984): 40–51.

Green, Harry W., II. "Solving the Paradox of Deep Earthquakes." *Scientific American* 271 (September 1994): 64–71.

Gurnis, Michael. "Ridge Spreading, Subduction, and Sea Level Fluctuations." *Science* 150 (November 16, 1990): 970–972.

Kerr, Richard A. "Having It Both Ways in the Mantle." *Science* 258 (December 1992): 1576–1578.

Macdonald, Kenneth C. and Paul J. Fox. "The Mid-Ocean Ridge." *Scientific American* 262 (June 1990): 72–79.

Monastersky, Richard. "Mid-Atlantic Ridge Survey Hits Bull's-Eye." *Science News* 135 (May 13, 1989): 295.

Peacock, Simon M. "Fluid Processes in Subduction Zones." *Science* 248 (April 20, 1990): 329–336.

Powell, Corey S. "Peering Inward." *Scientific American* 264 (June 1991): 100–111.

Sullivan, Walter. "Earth's Crust Sinks Deep, Only to Rise in Plumes of Lava." *The New York Times* (June 15, 1993): C1 & C8.

Wickelgren, Ingrid. "Simmering Planet." *Discover* 11 (July 1990): 73–75.

Zimmer, Carl. "The Ocean Within." *Discover* 15 (October 1994): 20–21.

SUBMARINE VOLCANOES

Berreby, David. "Barry Versus the Volcano." *Discover* 12 (June 1991): 61–67.

Coffin, Millard F. and Olav Eldholm. "Large Igneous Provinces." *Scientific American* 269 (October 1993): 42–49.

Courtillot, Vincent E. "A Volcanic Eruption." *Scientific American* 263 (October 1990): 85–92.

Dvorak, John J., Carl Johnson, and Robert I. Tilling. "Dynamics of Kilauea Volcano." *Scientific American* 267 (August 1992): 46–53.

Kerr, Richard A. "Did Pinatubo Send Climate-Warming Gases into a Dither?" *Science* 263 (March 18, 1994): 1562.

Lewis, G. Brad. "Island of Fire." *Earth* 4 (October 1995): 32–33.

Monastersky, Richard. "Garden of Volcanoes in the Pacific." *Science News* 143 (June 5, 1993): 367.

Oliwenstein, Lori. "Lava and Ice" *Discover* 13 (October 1992): 18.

Rank, David. "Seeing Spots." *Earth* 7 (February 1998): 18–19.

Richardson, Randall M. "Bermuda Stretches a Point." *Nature* 350 (April 25, 1991): 655.

Vink, Gregory E., W. Jason Morgan, and Peter R. Vogt. "The Earth's Hot Spots." *Scientific American* 252 (April 1985): 50–57.

White, Robert S. and Dan P. McKenzie. "Volcanism at Rifts." *Scientific American* 261 (July 1989): 62–71.

ABYSSAL CURRENTS

Broeker, Wallace S. "Chaotic Climate." *Scientific American* 262 (November 1995): 62–68.

Brown, Kathryn. "The Motion of the Ocean." *Science News* 158 (July 15, 2000): 42–44.

D'Agnese, Joseph. "Why Has Our Weather Gone Wild?" *Discover* 21 (June 2000): 72–78.

Folger, Tim. "Waves of Destruction." *Discover* 15 (May 1994): 68–73.

Garrett, Chris. "A Stirring Tale of Mixing." *Nature* 364 (August 19, 1993): 670–671.

Gonzalez, Frank I. "Tsunami!" *Scientific American* 280 (May 1999): 56–65.

Kerr, Richard A. "Big El Niños Ride the Back of Slower Climate Change." *Science* 283 (February 19, 1999): 1108–1109.

———. "Ocean-in-a-Machine Starts Looking Like the Real Thing." *Science* 260 (April 2, 1993): 32–33.

Kunzig, Robert. "The Iron Man's Revenge." *Discover* 15 (June 1994): 32–35.

Lockridge, Patricia A. "Volcanoes and Tsunamis." *Earth Science* 42 (Spring 1989): 24–25.

Monastersky, Richard. "Getting the Drift of Ocean Circulation." *Science News* 144 (August 21, 1993): 117.

Pendick, Daniel. "Waves of Destruction." *Earth* 6 (February 1997): 27–29.

Zimmer, Carl. "The North Atlantic Cycle." *Discover* 16 (January 1995): 77.

COASTAL GEOLOGY

Friedman, Gerald M. "Slides and Slumps." *Earth Science* 41 (Fall 1988): 21–23.

Holloway, Marguerite. "Soiled Shores." *Scientific American* 265 (October 1991): 103–116.

Horgan, John. "The Big Thaw." *Scientific American* 274 (November 1995): 18–20.

———. "Antarctic Meltdown." *Scientific American* 268 (March 1993): 19–28.

Maslin, Mark. "Waiting for the Polar Meltdown." *New Scientist* 139 (September 4, 1993): 36–41.

Monastersky, Richard. "Against the Tide." *Science News* 156 (July 24, 1999): 63.

Noris, Robert M. "Sea Cliff Erosion: A Major Dilemma." *Geotimes* 35 (November 1990): 16–17.

Parfit, Michael. "Polar Meltdown." *Discover* 10 (September 1989): 39–47.

Peltier, W. R. "Global Sea Level and Earth Rotation." *Science* 240 (May 13, 1988): 895–900.

Pnenvenne, Laura Jean. "The Disappearing Delta." *Earth* 5 (August 1996): 16–17.

Schaefer, Stephen J. and Stanley N. Williams. "Landslide Hazards." *Geotimes* 36 (May 1991): 20–22.

Zimmer, Carl. "Landslide Victory." *Discover* 12 (February 1991): 66–69.

SEA RICHES

Barnes, H. L. and A. W. Rose. "Origins of Hydrothermal Ores." *Science* 279 (March 27, 1998): 2064–2065.

Borgese, Elisabeth Mann. "The Law of the Sea." *Scientific American* 248 (March 1983): 42–49.

Brimhall, George. "The Genesis of Ores." *Scientific American* 264 (May 1991): 84–91.

Carpenter, Betsy. "Opening the Last Frontier." *U.S. News and World Report* 105 (October 24, 1988): 64–66.

Davis, Ged R. "Energy for Planet Earth." *Scientific American* 263 (September 1990): 55–62.

Fulkerson, William, Roddie R. Judkins, and Manoj K. Sanghvi. "Energy from Fossil Fuels." *Scientific American* 263 (September 1990): 129–135.

Hapgood, Fred. "The Quest for Oil." *National Geographic* 176 (August 1989): 226–263.

Penney, Terry R. and Desikan Bharathan. "Power from the Sea." *Scientific American* 256 (January 1987): 86–92.

Rosenberg, A. A., et al. "Achieving Sustainable Use of Renewable Resources." *Science* 262 (November 5, 1993): 828–829.

Suess, Erwin, et. al. "Flammable Ice." *Scientific American* 281 (November 1999): 76–83.

Weiss, Peter. "Oceans of Electricity." *Science News* 159 (April 14, 2001): 234–236.

MARINE BIOLOGY

Brown, Barbara E. and John C. Ogden "Coral Bleaching." *Scientific American* 268 (January 1993): 64–70.

Chen, Ingfei. "Great Barrier Reef: A Youngster to the Core." *Science* 138 (December 8, 1990): 367.

Culotta, Elizabeth. "Is Marine Biodiversity at Risk?" *Science* 263 (February 18, 1994): 918–920.

Elmer-Dewitt, Philip. "Are Sharks Becoming Extinct?" *Time* 137 (March 4, 1991): 67.

Fritz, Sandy. "The Living Reef." *Popular Science* 246 (May 1995): 48–51.

Glausiusz, Josie. "The Old Fish of the Sea." *Discover* 23 (January 1999): 49.

Kerr, Richard A. "From One Coral Many Findings Blossom." *Science* 248 (June 15, 1990): 1314.

Monastersky, Richard. "The Whale's Tale." *Science News* 156 (November 6, 1999): 296–298.

Perkins, Sid. "The Latest Pisces of an Evolutionary Puzzle." *Science News* 159 (May 5, 2001): 282–284.

Schreeve, James. "Are Algae—Not Coral—Reefs' Master Builders?" *Science* 271 (February 2, 1996): 597–598.

Simpson, Sarah, "Looking for Life Below." *Scientific American* 282 (June 2000): 94–101.

Stover, Dawn. "Creatures of the Thermal Vents." *Popular Science* 246 (May 1995): 55–57.

Svitil, Kathy A. "It's Alive, and It's a Graptolite." *Discover* 14 (July 1993): 18–19.

RARE SEAFLOOR FORMATIONS

Alvarez, Walter and Frank Asaro. "An Extraterrestrial Impact." *Scientific American* 263 (October 1990): 78–84.

Broad, William J. "A Voyage Into the Abyss: Gloom, Gold, and Godzilla." *The New York Times* (November 2, 1993): C1 & C12.

Fischman, Josh. "In Search of the Elusive Megaplume." *Discover* 20 (March 1999): 108–115.

Fryer, Patricia. "Mud Volcanoes of the Marianas." *Scientific American* 266 (February 1992): 46–52.

Gorman, Christine. "Subterranean Secrets." *Time* 140 (November 30, 1992): 64–67.

Grieve, Richard A. "Impact Cratering on the Earth." *Scientific American* 262 (April 1990): 66–73.

Kerr, Richard A. "Testing an Ancient Impact's Punch." *Science* 263 (March 11, 1994): 1371–1372.

———. "Surging Plate Tectonics Erupts Beneath the Sea." *Science* 250 (December 1990): 21.

Lipske, Mike. "Wonder Holes." *International Wildlife* 20 (January/February 1990): 47–51.

Monastersky, Richard. "The Light at the Bottom of the Ocean." *Science News* 150 (September 7, 1996): 156–157.

———. "When Kilauea Crumbles." *Science News* 147 (April 8, 1995): 216–218.

Schueller, Gretel. "One-Note Eruptions." *Earth* 7 (February 1998): 14.

INDEX

Boldface page numbers indicate extensive treatment of a topic. *Italic* page numbers indicate illustrations or captions. Page numbers followed by *m* indicate maps; *t* indicate tables; *g* indicate glossary.